Lecture Notes in Computer Science　　　9689

Commenced Publication in 1973
Founding and Former Series Editors:
Gerhard Goos, Juris Hartmanis, and Jan van Leeuwen

More information about this series at http://www.springer.com/series/7410

François-Xavier Standaert · Elisabeth Oswald (Eds.)

Constructive Side-Channel Analysis and Secure Design

7th International Workshop, COSADE 2016
Graz, Austria, April 14–15, 2016
Revised Selected Papers

Springer

Editors
François-Xavier Standaert
UCL Crypto Group
Louvain-la-Neuve
Belgium

Elisabeth Oswald
University of Bristol
Bristol
UK

ISSN 0302-9743   ISSN 1611-3349 (electronic)
Lecture Notes in Computer Science
ISBN 978-3-319-43282-3   ISBN 978-3-319-43283-0 (eBook)
DOI 10.1007/978-3-319-43283-0

Library of Congress Control Number: 2016945799

LNCS Sublibrary: SL4 – Security and Cryptology

Printed on acid-free paper

This Springer imprint is published by Springer Nature
The registered company is Springer International Publishing AG Switzerland

Preface

The 7th International Workshop on Constructive Side-Channel Analysis and Secure Design (COSADE) was held in Graz, Austria, during April 14–15, 2016. This now well-established workshop brings together researchers from academia, industry, and government who share a common interest in the design and secure implementation of cryptographic primitives. COSADE 2016 received 32 submission; the review process relied on the EasyChair system.

From the pool of submissions, 12 high-quality papers were selected carefully after deliberations of the 30 Program Committee members who were supported by 24 additional reviewers. The composition of the Program Committee was representative of the good mix between academic and industrial researchers as well as the geographic spread of researchers across the globe. We would like to express our sincere gratitude to both the Program Committee members and reviewers.

As it has become custom, the Program Committee members voted on the best paper among the accepted papers. The resulting winner was "Exploiting the Physical Disparity: Side-Channel Attacks on Memory Encryption" authored by Thomas Unterluggauer and Stefan Mangard. The program also featured three invited talks. Tom Chothia elaborated on advanced statistical tests for detecting information leakage. François Dupressoir spoke about formal and compositional proofs of probing security for masked algorithms. Aurélien Francillon discussed what security problems can be spotted with large-scale static analysis of systems. We would like to thank the invited speakers for joining us in Graz.

Finally, we would like to thank the local organizers, in particular Stefan Mangard (general chair) and Thomas Korak, for their support and for making this great event possible. On behalf of the COSADE community we would also like to thank our GOLD sponsors Infineon Technologies AG, NewAE Technology Inc., NXP Semiconductors, Riscure, and Secure-IC, as well as our SILVER sponsors Rambus Cryptography Research and Oberthur Technologies, for their support.

And most importantly, we would like to thank the authors for their excellent contributions.

May 2016

Elisabeth Oswald
François-Xavier Standaert

Organization

Program Committee

Josep Balasch — KU Leuven, Belgium
Guido Bertoni — STMicroelectronics, Italy
Shivam Bhasin — Nanyang Technological University, Singapore
Christophe Clavier — University of Limoges, France
Hermann Drexler — Giesecke & Devrient, Germany
Cécile Dumas — CEA LETI, France
Thomas Eisenbarth — WPI, USA
Wieland Fischer — Infineon Technologies, Germany
Benoît Gérard — DGA Maîtrise de l'Information, France
Christophe Giraud — Oberthur Technologies, France
Vincent Grosso — UCL, Belgium
Johann Groszschädl — University of Luxembourg, Luxembourg
Tim Güneysu — University of Bremen, Germany
Sylvain Guilley — Télécom ParisTech, France
Johann Heyszl — Fraunhofer AISEC, Germany
Naofumi Homma — Tohoku University, Japan
Michael Hutter — CRI, USA
Ilya Kizhvatov — Riscure, The Nederlands
Thanh-ha Le — Morpho, France
Kerstin Lemke-Rust — Bonn-Rhein-Sieg University of Applied Sciences, Germany
Marcel Medwed — NXP Semiconductors, Austria
Amir Moradi — Ruhr-Universität Bochum, Germany
Debdeep Mukhopadhyay — Indian Institute of Technology Kharagpur, India
Elisabeth Oswald — University of Bristol, UK
Emmanuel Prouff — ANSSI, France
Francesco Regazzoni — University of Lugano, Switzerland
Matthieu Rivain — CryptoExperts, France
Kazuo Sakiyama — The University of Electro-Communications Tokyo, Japan
Francois-Xavier Standaert — UCL Crypto Group, Belgium
Carolyn Whitnall — University of Bristol, UK

Additional Reviewers

Abdullin, Nikita
Barbu, Guillaume
Bauer, Sven
Becker, Georg T.
Bocktaels, Yves
Breier, Jakub
Chabrier, Thomas
Chen, Cong
Dabosville, Guillaume
De Santis, Fabrizio
Dinu, Daniel
Goodwill, Gilbert
Greuet, Aurélien
Hayashi, Yuichi

He, Wei
Hoffmann, Lars
Irazoqui, Gorka
Jap, Dirmanto
Knezevic, Miroslav
Li, Yang
Lomne, Victor
Longo Galea, Jake
Martin, Daniel
Mather, Luke
Melzani, Filippo
Miura, Noriyuki
Oder, Tobias
Omic, Jasmina

Patranabis, Sikhar
Riou, Sebastien
Samarin, Peter
Sasdrich, Pascal
Schellenberg, Falk
Schneider, Tobias
Selmke, Bodo
Susella, Ruggero
Takahashi, Junko
Ueno, Rei
Vermoen, Dennis
Yli-Mayry, Ville

Contents

Side-Channel Analysis (Tools)

Security and Physical Attacks

Exploiting the Physical Disparity: Side-Channel Attacks on Memory Encryption

Thomas Unterluggauer[(✉)] and Stefan Mangard

Institute for Applied Information Processing and Communications,
Graz University of Technology, Inffeldgasse 16a, 8010 Graz, Austria
{thomas.unterluggauer,stefan.mangard}@iaik.tugraz.at

Abstract. Memory and disk encryption is a common measure to protect sensitive information in memory from adversaries with physical access. However, physical access also comes with the risk of physical attacks. As these may pose a threat to memory confidentiality, this paper investigates contemporary memory and disk encryption schemes and their implementations with respect to Differential Power Analysis (DPA) and Differential Fault Analysis (DFA). It shows that DPA and DFA recover the keys of all the investigated schemes, including the tweakable block ciphers XEX and XTS. This paper also verifies the feasibility of such attacks in practice. Using the EM side channel, a DPA on the disk encryption employed within the ext4 file system is shown to reveal the used master key on a Zynq Z-7010 system on chip. The results suggest that memory and disk encryption secure against physical attackers is at least four times more expensive.

Keywords: Memory encryption · Side-channel attack · Power analysis · DPA · Fault analysis · DFA · Ext4

1 Introduction

Many electronic computing devices nowadays contain and process sensitive data in hostile environments. Among two particularly relevant examples, the first are engineering companies whose production machines are shipped around the world. These machines contain high-value intellectual property, e.g., control parameters and source code, that their vendors wish to be protected from unauthorized access and proliferation. Similarly, malicious access and modification must be prevented if usage statistics are used for billing. The second example are employee smart phones or laptops containing corporate secrets. Unattended such devices are a highly interesting target for industrial espionage and therefore need protection mechanisms.

In both examples, adversaries interested in the sensitive data potentially have physical access to the device. To prevent these attackers from simply reading confidential information from main or external memory, e.g., hard disks and memory cards, encryption of memory is well established. Several dedicated encryption

© Springer International Publishing Switzerland 2016
F.-X. Standaert and E. Oswald (Eds.): COSADE 2016, LNCS 9689, pp. 3–18, 2016.
DOI: 10.1007/978-3-319-43283-0_1

modes for memory, such as Cipher-Block-Chaining with Encrypted Salt-Sector IV (CBC-ESSIV) [10], Xor-Encrypt-Xor (XEX) [25], and XEX-based Tweaked codebook mode with ciphertext Stealing (XTS) [1], were proposed to fulfill the special requirements of memory encryption. These successfully prevent a variety of attacks, ranging from simple dumps of memory cards or hard disks to bus probing and cold boot attacks [14], and are thus implemented in an increasing number of real-world applications, such as dm-crypt, Mac OS X, Android, and ext4.

However, one important aspect contemporary memory encryption schemes left unconsidered are physical attacks such as side-channel and fault attacks. These allow the adversary to learn about secret key material used during encryption from various side channels, e.g., power, timing, and Electromagnetic Emanation (EM), or from faulty computations due to intentionally induced faults, e.g., clock glitches. Given physical access of the adversary as the motivating threat for memory encryption, physical attacks must not be neglected as these would allow adversaries to learn the encryption key and thus to decrypt confidential data in memory. The consideration of physical attacks is particularly important for permanently running devices that are threatened by attackers without any time constraints, e.g., a corporate customer may be interested in the data and IP of an embedded control unit within a purchased production machine.

This paper therefore investigates contemporary memory encryption schemes and their implementation within dm-crypt, Android 5.0, Mac OS X and ext4 in terms of physical attacks. As one main result, our detailed analysis shows that Differential Power Analysis (DPA) and Differential Fault Analysis (DFA) breaks all contemporary memory and disk encryption schemes used in practice. Most prominently, it presents tricks to be applied to DPA and DFA in order to obtain the keys from the tweakable ciphers XEX and XTS. Supporting the analysis results, our second contribution exploits the EM side channel of a Zynq-7010 system on chip in a practical attack on the recently introduced ext4 disk encryption mechanism that completely discloses the confidential disk content. We thus conclude that securing memories against physical adversaries by using contemporary memory encryption requires protected implementations, e.g., [5, 16,22], that increase the cost of memory encryption at least by a factor of four.

This paper is organized as follows. Section 2 introduces memory encryption and gives an overview on common state-of-the-art implementations. The memory encryption schemes are analyzed with respect to both DPA and DFA in Sect. 3. The practical feasibility of such attacks is evaluated in Sect. 4, and Sect. 5 concludes the paper.

2   Memory Encryption

Memory encryption deals with the encryption of data contained in memory such as RAM, memory cards and hard disks. However, in practice different variants and notations are being used for memory encryption. This Section therefore defines memory encryption and gives an overview on common memory encryption schemes and implementations.

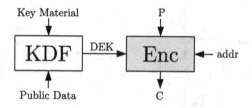

Fig. 1. Generic model of memory encryption.

2.1 Definition

The encryption of memory is usually performed using dedicated memory encryption schemes as these schemes have to fulfill several requirements: (1) ensure random access to all memory blocks, (2) provide sufficiently fast bulk encryption, and (3) the only information an adversary can derive from the encrypted memory is whether a memory block has changed or not.

Definition 1. *A memory encryption scheme is an encryption scheme Enc* : $\mathcal{K} \times \mathcal{A} \times \{0,1\}^n \to \{0,1\}^n$, *which*

- *uses a key K from key space $\mathcal{K}$, and*
- *splits the memory into $s = \lceil \frac{size_{memory}}{n} \rceil$ n-bit memory blocks,*
- *identifies each of the memory blocks by their address in address space $\mathcal{A}$, and*
- *provides address-dependent en-/decryption for each of these memory blocks.*

Definition 1 considers the encryption of a flat memory space and requires the encryption process to incorporate address information. The address information allows memory encryption schemes to fulfill requirement (3) as for this reason each memory block is encrypted differently. Otherwise, it would be easily recognizable if certain data is contained in different memory locations and valid (but encrypted) data could simply be copied to different addresses (*splicing attack* [9]). The requirements (1) and (2) are typically satisfied by splitting the memory space into blocks using two different granularities: the memory is divided into larger sectors (or pages) and each sector (or page) is divided into encryption blocks. The encryption mode then ensures fast bulk encryption within each sector and random access on sector level.

2.2 Memory Encryption in Practice

In practice, memory encryption is often named disk encryption referring to the type of memory used. There are two variants of disk encryption: (1) *block device* or *full disk encryption*, and (2) *file-level disk encryption*. While *full disk encryption* performs encryption directly on the raw memory space of a whole disk, block device, or partition, i.e., beneath a file system, *file-level disk encryption* performs encryption on file level on top of or within a file system. Both variants use the same sort of memory encryption schemes, but apply them to different

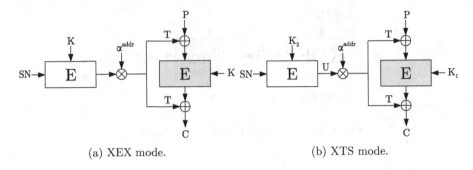

(a) XEX mode.   (b) XTS mode.

Fig. 2. Tweakable ciphers for disk encryption.

portions of the memory. Throughout the paper, the term memory encryption thus denotes any of these variants.

Another aspect of practical implementations of memory encryption is that they usually employ a Key Derivation Function (KDF) to derive the Data Encryption Key (DEK) to be used within the memory encryption scheme from, e.g., a user password and public nonces. The combination of such a KDF and a memory encryption scheme leads to the generic model of memory encryption in Fig. 1. The following will use this model to first describe typical schemes for both the KDF and the encryption part, and will then show how these are used in several practical implementations.

Key Derivation Functions. To derive a key from a user password or a PIN, typically a password hashing function such as PBKDF2 [18] or scrypt [23] is used. This password-derived key is then mostly used as a Key Encryption Key (KEK) to decrypt the actual master key MK of the memory using an ordinary block cipher. Depending on the concrete setup, such master key MK is directly used as the DEK for the memory encryption scheme or is used to further derive or decrypt keys, e.g., DEKs for the encryption of single files in file-level disk encryption.

Encryption Schemes. Common implementations exclusively deal with the encryption of external memory, e.g., hard disks. These implementations, e.g., in dm-crypt, mainly utilize the modes XEX [25], XTS [1], and CBC with ESSIV [10]. The tweakable block ciphers XEX and XTS are shown in Fig. 2. Both encryption modes apply a tweak T to the cipher E that results from a binary-field multiplication of the encrypted sector number with the memory block address. While XEX uses only one key, XTS uses two different keys for the two instances of the cipher. The CBC mode with Encrypted Salt-Sector IV (ESSIV) is depicted in Fig. 3. ESSIV ensures a secret IV and thus prevents watermarking attacks [27]. It computes the IV as the encryption of the sector number with the hashed key (i.e., salt).

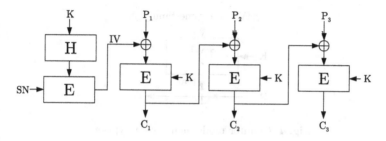

Fig. 3. Disk encryption via CBC and ESSIV.

Differently, research on the design and construction of secure systems further considered the encryption of the main memory. Primarily variants of the counter mode encryption were proposed such as in Fig. 4 [26,29]. The pad is the encryption of a block-specific seed that comprises an Initial Vector (IV), the memory block address, and a timestamp (or counter). It is mostly favored due to the little latency it introduces on the path to the memory.

2.3 State-of-the-Art Implementations

The following presents common implementations within dm-crypt, Android, Mac OS X, and ext4, and shows that the memory encryption schemes presented before have high prevalence throughout all of these implementations.

dm-crypt. dm-crypt [2] is a disk encryption utility that provides transparent encryption of arbitrary block devices within Linux $\geq$ 2.6, i.e., block device encryption. dm-crypt can be configured to use one of several available encryption modes, i.a., CBC-ESSIV and XTS (default), using different block ciphers, e.g., AES-128 [8]. The utility requires the user to supply the block device DEK when mounting the block device. For more convenient usage, however, Linux Unified Key Setup (LUKS) [11] can be used. LUKS adds a meta-data header to the block device that stores the encrypted DEK. The respective KEK is derived from a user password using PBKDF2.

Mac OS X. Mac OS X from version 10.7 (Lion) onwards provides block device encryption using the tool FileVault 2 [3,7]. Mac OS X encrypts block devices using XTS and AES-128 with separate DEKs that are chosen randomly upon setup of each encrypted block device. For key storage, Mac OS X uses a three-tier hierarchy of DEKs, KEKs and Derived Key Encryption Keys (DKEK). The DEK is encrypted using a randomly chosen KEK that is encrypted using at least one DKEK. DKEKs can, e.g., be derived from a password or be the public key of a corporate certificate. Both the DEK and the KEK are stored encrypted in a meta-data block on the block device.

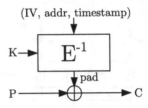

Fig. 4. Counter mode memory encryption.

Android. Android is equipped with full disk encryption for devices such as flash memory. In Android 5.0, encryption of block devices is based on `dm-crypt` that is configured to use AES-128 and CBC-ESSIV [13]. Its DEK is sized 128 bits by default and stored encrypted on the block device. The respective KEK is derived from a user password and a hardware-bound key using scrypt and a signing procedure within a Trusted Execution Environment (TEE).

Ext4. Since Linux 4.1, the ext4 file system offers file-level disk encryption [20,21]. It allows to set up encryption for a specific folder that is assigned a master key derived from a user passphrase and a salt using PBKDF2. While ext4 encrypts file content and names, meta data and file system structure is available in plaintext. Each file uses an individual DEK that is derived from the master key MK and a file nonce N_f using AES-128 in ECB mode, i.e., $DEK_f = E_{N_f}(MK)$. The respective nonce N_f is stored in the file's meta-data section. The file DEK is used to encrypt the file contents using XTS and AES-128.

3   Physical Attacks on Memory Encryption

Physical access as the motivation for memory encryption and the prevalence of the memory encryption schemes from Sect. 2 necessitate their analysis with respect to physical attacks such as side-channel and fault attacks. The following analysis of memory encryption schemes w.r.t. physical attacks shows that both DPA [19] and DFA [4] attacks are easily capable of breaking all the schemes presented, i.e., they reveal the DEK that allows to decrypt all memory content. Most remarkably, it demonstrates how to obtain the AES-128 keys in the tweakable block ciphers XEX and XTS with practical complexity.

3.1   Differential Power Analysis

DPA attacks and its variants, e.g., Correlation Power Analysis [6], are methods that allow recovery of an encryption key based on power measurements or similar, e.g., EM. The typical procedure is to measure the power of n different en-/decryptions for known plain- or ciphertext, to compute certain intermediate values within the en-/decryption based on the n different plain-/ciphertexts and the possible keys, and to map the intermediate values to hypothetical power consumptions according to a leakage model. Correlation of the power traces of the n en-/decryptions with the respective hypothetical power consumptions reduces the key space or determines the key uniquely. The following details DPA attack scenarios on the schemes from Sect. 2.

XEX Mode. The tweak T makes sure that the block cipher behaves differently for each memory address. In spite of this, DPA-style attacks are applicable with little modifications. Therefore, the adversary focuses on one particular memory block, i.e., fixed sector and fixed memory address. For this memory block, the adversary observes ciphertexts and power traces of several encryption processes. The captured power traces are then used twice to attack different rounds of the block cipher shaded gray in Fig. 2a, as the following illustrates for AES-128:

1. From an attacker's point of view, the last round key rk_{10} is blinded with the tweak T. However, for a fixed sector and memory address, the tweak T is constant. A DPA that targets the input of the last round's SBox will thus reveal the last round key xored with the tweak, i.e., $rk_{10} \oplus T$.
2. Knowledge of $rk_{10} \oplus T$ is sufficient to target the input of the second-last round's SBox in a second DPA. It reveals the second-last round key rk_9, which can be used to compute the key K.

Two consecutive DPAs on the same set of traces allow to gain knowledge of the key K. The DPAs disclose the information contained in all memory blocks across all sectors, even though only one particular block in one specific sector is actually attacked.

Note that besides standard DPA, also unknown plaintext template attacks [15] are applicable to directly obtain rk_{10}. However, such attacks require a preceding profiling step to create suitable templates. Alternatively, if the adversary additionally has knowledge of the accessed sector, e.g., from the observation of memory addresses on the bus, the attack generally becomes easier. In this case, the tweak computation that encrypts the sector number can be attacked to immediately learn K from power traces of memory accesses to different sectors. However, depending on the practical circumstances, either of those attacks is more suitable, e.g., the adversary may want to avoid raising suspicion by not probing the memory bus.

XTS Mode. Contrary to XEX, a successful DPA on XTS requires the knowledge of the accessed sector number. It allows to first obtain K_2 from the tweak computation. Once K_2 is known, the tweak T used for encrypting any memory block can be computed which enables a straight-forward attack on the key K_1 by monitoring the power consumption during arbitrary memory accesses.

Counter Mode. Known-plaintext scenarios allow for DPA attacks that recover the key K in counter mode encryption. They facilitate the computation of the encryption pads from both known plain- and ciphertexts and thus DPA on the last round of the cipher. Typically, plaintexts would be assumed to be unknown since memory encryption is applied. However, known-plaintext scenarios will certainly occur in memory encryption. One such case would be publicly known (or observable) data that is sent to a device, e.g., via external interfaces, and that is consecutively encrypted and stored in main memory, e.g., within an input buffer.

If there are insufficiently many known plaintexts, a known input seed also allows for a DPA - one that does not even require any ciphertext. Often, the counters and addresses within the seed will be publicly accessible (or observable). If the IV is public as well, the seed will be fully known and a DPA in the first round of the cipher be possible. The IV will mostly be stored publicly on the disk for disk encryption, but might be chosen randomly at startup and remain inaccessible for encryption of the main memory. Still, the approach in [17], where a DPA is performed on the counter mode of AES without knowledge of the counter value, might be applicable.

CBC Mode with ESSIV. Independently of the initial vector derivation, DPA attacks on the CBC mode are trivially possible through the observation of ciphertexts and power traces of the respective encryption processes. The recovered key K then allows to compute each sector's IV (ESSIV) and hence to obtain any plaintext.

3.2   Differential Fault Analysis

Differential Fault Analysis (DFA) [4] describes techniques that use algebraic properties of ciphers to find out about the key from one correct and one or several faulty cipher invocations with the same input. Various techniques to inject faults into a device exist, e.g., power and clock glitches, laser shots, and electromagnetic pulses. However, the following investigation does not consider how the faults are injected, but elaborates on how faults are exploited in order to obtain the key. It details DFA attack scenarios on the schemes from Sect. 2, and most noteworthy, how to break the tweakable block ciphers XEX and XTS with practical complexity 2^{35} if AES-128 is used.

XEX Mode. The attack procedure of DFA to learn the key K is tightly linked with the employed cipher. Exemplarily, we show how to use DFA to extract the key from AES-128 in XEX mode. The DFA targets the block cipher that is shaded gray in Fig. 2a and consists of two basic steps:

1. An arbitrary byte fault in round 8 is used to extract the xor of round key 10 and the tweak ($rk_{10} \oplus T$).
2. A byte fault in round 7 and a modified representation of the AES round function lead to round key 9 and thus the key K.

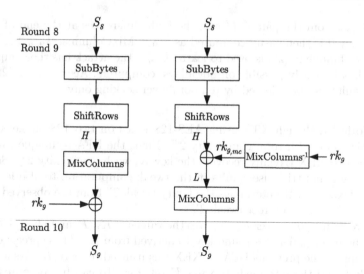

Fig. 5. AES round function (left) and its alternative representation (right).

Learning $rk\_10 \oplus T$. From an attacker's point of view, the last round key rk_10 is blinded with the tweak T. This requires the tweak T to be constant for DFA, i.e., the attack operates on fixed sector and fixed memory block. By forcing reencryption of the same plaintext in the desired block, the adversary gets the chance to inject an arbitrary byte fault during round 8 of the encryption process of the tweakable cipher. Application of a suitable DFA technique, e.g., [24, 28], to the pair of right and faulty ciphertext results in the value $rk\_10 \oplus T$.

Learning round key 9. The DFA to learn rk_9 benefits from an alternative representation of the AES round function. As shown in Fig. 5, it is obtained from swapping MixColumns and AddRoundKey. The linearity of MixColumns allows this transformation if the round key is modified accordingly, i.e.,

$$\text{MixColumns}(H) \oplus rk_9 = \text{MixColumns}(H \oplus \text{MixColumns}^{-1}(rk_9))$$
$$= \text{MixColumns}(H \oplus rk_{9,mc}).$$

In the following, the alternative representation of the round function is used for round 9. The attack starts by injecting a random byte fault during round 7. As the MixColumns operation propagates the fault to the other state bytes, all bytes are affected by the end of round 8. The observed pair of right and faulty ciphertext C, C' and the value $rk_{10} \oplus T$ are used to compute backward to obtain the respective values L, L' in round 9.

Interpreting L, L' as a pair of right and faulty ciphertext, the remaining cipher looks like a round-reduced version of the AES with one inner round missing. The last round consists of AddRoundKey, ShiftRows, and SubBytes and uses the round key $rk_{9,mc}$. The benefit of this approach is that now any DFA technique that targets the last round key of the AES, e.g., [24, 28], is suitable to obtain $rk_{9,mc}$ from the pair L, L' and the fault differences at the end of round 8. Round key 9 is then easily computed as $rk_9 = \text{MixColumns}(rk_{9,mc})$.

If the technique in [28] is used to learn $rk_{9,mc}$, the attack has the complexity 2^{34} and thus is clearly possible on nowadays' computers. According to [28], the required faults can be injected by temporal overclocking only.

XTS Mode. Although XTS using AES-128 relies on two 128-bit keys, DFA breaks this mode with total complexity 2^{35}. First, the DFA technique that was just applied to XEX trivially recovers the key K_1 with complexity 2^{34}. Second, the following small trick uses faults in the tweak computation to also learn K_2 with complexity 2^{34}. It determines the faulty tweak T' from the observed faulty ciphertext C' and the correct tweak T.

The procedure to recover K_2 requires the values of K_1, P, and $rk_{1,10} \oplus T$ to be known, where $rk_{1,10}$ denotes round key 10 derived from K_1. These preconditions usually apply if the previous DFA on XEX was utilized to learn K_1. As a result, the tweak T and the intermediate value U (cf. Fig. 2b) can be computed: $U = \alpha^{-addr} \cdot T$. A random fault that is injected in one byte of the state in round 9 of the AES affects four bytes of U. Although the respective faulty U' is not directly observable, it can be brute-forced with complexity 2^{32}. This is done by trying all values for the faulty bytes of U', computing the respective tweaks T', encrypting the original plaintext P using T' and K_1, and matching the result against the faulty ciphertext C'. Once U' is known, four bytes of $rk_{2,10}$ (round key 10 derived from K_2) are revealed using the technique in [24]. Hereby, the possible key space for $rk_{2,10}$ is reduced by the possible differences that can be observed at the output of MixColumns in round 9 that result from a single byte fault during round 9. Similarly, three more faults in different bytes of the state of round 9 recover the remaining 12 bytes of $rk_{2,10}$ and thus K_2.

Counter Mode. DFA on a block cipher operated in counter mode (cf. Fig. 4) requires access to the output of the cipher, i.e., the pad. Since encryption pads must not repeat, consecutive encryptions of plaintexts will not use the same pad and encryption seed. As a result, DFA is limited to the decryption process. If the same ciphertext is loaded from the same memory address several times and the adversary can inject faults during the pad computations and observe the respective plaintexts, the correct and faulty pads can be computed and the master key K be learned via a suitable DFA technique. The required plaintexts may be observed from communication of the device via external interfaces.

CBC Mode with ESSIV. Independently of the initial vector derivation, DFA is trivially possible by restricting analysis to one specific memory block within the CBC chain of one particular sector. Therefore, reencryption of the same plaintext has to be triggered for the desired memory block, e.g., through placing the same message in an input buffer by repeatedly sending the same message to the device. Faults injected during reencryption are directly observable in the resulting ciphertext. This facilitates the application of a suitable DFA technique in order to learn the master key K. Note that for this to work, all memory blocks in the sector prior to the target block must not change during reencryption.

4 EM Attack on Ext4 Encryption

As our analysis points out, contemporary memory encryption schemes are clearly vulnerable to physical attacks. However, it remains to show that such attacks are indeed feasible on contemporary systems. This Section therefore demonstrates a practical attack on the disk encryption scheme incorporated into the ext4 file system. The EM attack conducted on a Zynq Z-7010 system on chip (SoC) reveals the used master key and thus all content by exploiting the leakage of the first round of an AES execution.

4.1   Analysis of Ext4 Disk Encryption

Disk encryption within the ext4 file system works on file level and allows to encrypt arbitrary directories using a specified master key MK. For each file in such directory, the master key MK is used to derive an individual data encryption key DEK_f to encrypt the respective file's content and name. Key derivation is done by encrypting MK with AES-128 in ECB mode using a public file nonce N_f as the key. It starts whenever DEK_f is needed and not already present in main memory. The size of both MK and DEK_f is 512 bits and chosen such as to be able to encrypt files with AES-256 in XTS mode in future versions. However, currently only AES-128 in XTS mode is supported and thus the last 256 bits of DEK_f and MK are not used. The file nonce N_f is stored in an extended attribute of the file's inode.

Clearly, given the master key MK and a public file nonce N_f, the respective file key DEK_f can be derived. However, the key derivation chosen in ext4 also allows to compute the master key MK given any DEK_f and the respective nonce N_f. Therefore, an attacker who wants to learn MK using power analysis can choose between two equivalent targets, namely (1) data encryption of file content, and (2) the derivation of the file key DEK_f. In terms of target (1), the strategy from Sect. 3 can be straight-forwardly applied, but one may need files that are sufficiently large to be able to learn K_2 within XTS. With respect to target (2), one needs to monitor accesses to many different files as such trigger key derivations. To practically verify the feasibility of attacks on disk encryption, we opted for target (2).

4.2   General Attack Flow

The attack we performed assumes an encrypted folder on an SD card using the ext4 file system. It further assumes the attacker is able to trigger the creation of new files within the encrypted folder via external interfaces, e.g., by uploading data via a running web server or writing log files.

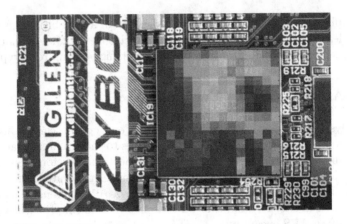

Fig. 6. Distribution of t-test results on the chip surface.

To perform the attack, the attacker first dumps the (encrypted) content of the SD card. They may not be able to read the actual content from such file system dump, but can learn about the directory structure as meta data is not encrypted. Second, the attacker triggers the creation of sufficiently many files on the SD card, observes the EM side channel, and stores the respective EM traces. Third, the attacker again dumps the content of the SD card. By comparing its content with the initial dump from before the measurements, the attacker can learn which files have been created. The meta data of the newly created files allows to both learn the used nonces N_f and their creation date, which in turn allows to map the newly created files on the SD card to the EM traces. In the next step, the attacker creates the power model for the key derivation, i.e., $DEK_f = E_{N_f}(MK)$. Finally, the power model is matched with the EM traces to reveal the master key.

To investigate the encrypted directory in the file system, debugging and forensic tools are highly suitable. We used the tool `debugfs` to find new files in the file system and to learn their creation date and the respective nonces. Note that the access times are also available within the file system, which allows for the described attack also when monitoring arbitrary file accesses.

4.3   Experimental Setup and Results

The feasibility of the attack on ext4 encryption in Sect. 4.2 was verified using the Digilent ZYBO board. The board hosts a Xilinx Zynq Z-7010 SoC, 512 MB of DDR3 RAM, and several IO interfaces, i.a., an SD card slot. The Zynq Z-7010 SoC combines an Artix-7 FPGA and a state-of-the-art hard macro comprising a 650-MHz dual-core ARM Cortex-A9 processor, IO modules, and memory controllers. The measurement devices required to capture the EM traces involved a LeCroy WavePro 725Zi oscilloscope, a Langer RF B 3-2 magnetic field probe, and a Langer PA 303 pre-amplifier.

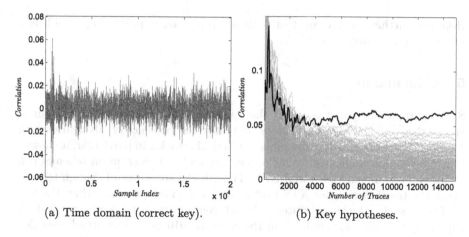

(a) Time domain (correct key).

(b) Key hypotheses.

Fig. 7. Single-byte correlation results for ext4 key derivation.

The general leakage behavior of the Zynq Z-7010 was examined by running the AES T-table implementation included in the Linux 4.3 kernel in a bare-metal application. Therefore, the EM probe was placed in different locations using a stepper table to evaluate a fixed vs. random t-test. This revealed the spots of high leakage as shown in Fig. 6 and allowed for successful DPA on the bare-metal AES.

The setup for the complete disk encryption scenario was established by configuring the Zynq SoC to use a 350-MHz memory clock and a 625-MHz CPU clock and deploying Linux 4.3 to the ZYBO board. An ext4 file system was created on an SD card and one directory encrypted such that it is only readable by the system running on the ZYBO board. The attack procedure from Sect. 4.2 was executed by repeatedly creating new files via the UART interface. The oscilloscope was triggered to capture an EM trace at 5 GS by setting a GPIO pin just before creating a new file. The SD card content was then analyzed on a PC using debugfs, the EM traces aligned, and a DPA performed on the SBox output of the first AES round using the Hamming Weight power model.

The results of the DPA on a single byte of the master key are given in Fig. 7. Using 15,000 EM traces, Fig. 7a clearly presents the correlation of the power model of the correct key guess in the time domain. Moreover, in Fig. 7b the correct key byte (black) is clearly distinguished from the remaining key hypotheses with 5,000 measurements.

In this feasibility study, the Linux kernel was reconfigured to omit symmetric multiprocessing, dynamic frequency scaling, and caches. Moreover, AES executions were highlighted in the captured EM traces through another hardware-triggered signal to help finding AES executions. This however does not affect the applicability of the attack. For example, [12] showed the practicality of attacking a free-running OpenSSL implementation of AES with active caches and frequency scaling on the TI Sitara platform that uses an ARM Cortex-A8.

However, further improvement of both setup and trace processing would definitely be interesting future work.

5 Conclusion

Summarizing, this paper unveiled that contemporary mechanisms that aim to ensure the confidentiality of memory content in the presence of adversaries with physical access are clearly vulnerable to physical attacks. In particular, it showed that all common implementations of memory and disk encryption schemes can easily be broken using DPA and DFA. The attacks are powerful enough to even break the tweakable cipher XTS that is most commonly used. Further, the feasibility of such attacks on state-of-the-art computing systems was verified by exploiting the EM side channel on the Zynq Z-7010 SoC. The attack revealed the master key of the disk encryption scheme incorporated into the ext4 file system and thus all encrypted content.

Our results suggest that if memory encryption is supposed to use current schemes in the future, cipher implementations with appropriate countermeasures must be used. However, the secure cipher implementations proposed so far were mainly designed for the use in embedded devices and might thus not yield the desired throughput for memory encryption. For example, the 1st-order threshold implementations in [5, 22] require 246 and 266 clock cycles for one AES execution, respectively. Additionally, these implementations add an area overhead of a factor of four that must hence also be expected for secure memory encryption based on such protected implementations. It thus remains future work to implement memory encryption that fulfills both the requirement for sufficient throughput and security against side-channel adversaries.

Acknowledgments. This work has been supported by the Austrian Research Promotion Agency (FFG) under grant number 845579 (MEMSEC).

References

1. IEEE Standard for Cryptographic Protection of Data on Block-Oriented Storage Devices. IEEE Std 1619-2007, April 2008
2. Dm-crypt: Linux Kernel Device-Mapper Crypto Target (2015). http://www.saout.de/misc/dm-crypt/
3. Apple Inc.: Apple Technical White Paper: Best Practices for Deploying FileVault 2 (2012)
4. Biham, E., Shamir, A.: Differential fault analysis of secret key cryptosystems. In: Kaliski Jr., B.S. (ed.) CRYPTO 1997. LNCS, vol. 1294, pp. 513–525. Springer, Heidelberg (1997)
5. Bilgin, B., Gierlichs, B., Nikova, S., Nikov, V., Rijmen, V.: A more efficient AES threshold implementation. In: Pointcheval, D., Vergnaud, D. (eds.) AFRICACRYPT 2014. LNCS, vol. 8469, pp. 267–284. Springer, Heidelberg (2014)

6. Brier, E., Clavier, C., Olivier, F.: Correlation power analysis with a leakage model. In: Joye, M., Quisquater, J.-J. (eds.) CHES 2004. LNCS, vol. 3156, pp. 16–29. Springer, Heidelberg (2004)

7. Choudary, O., Grobert, F., Metz, J.: Infiltrate the Vault: security analysis and decryption of lion full disk encryption. Cryptology ePrint Archive, report 2012/374 (2012). http://eprint.iacr.org/

8. Daemen, J., Rijmen, V.: The Design of Rijndael: AES-The Advanced Encryption Standard. Springer, Heidelberg (2002)

9. Elbaz, R., Champagne, D., Gebotys, C.H., Lee, R.B., Potlapally, N.R., Torres, L.: Hardware mechanisms for memory authentication: a survey of existing techniques and engines. Trans. Comput. Sci. **4**, 1–22 (2009)

10. Fruhwirth, C.: New methods in hard disk encryption. Technical report (2005)

11. Fruhwirth, C.: LUKS On-Disk Format Specification (2011). https://gitlab.com/cryptsetup/cryptsetup/wikis/LUKS-standard/on-disk-format.pdf

12. Longo, J., De Mulder, E., Page, D., Tunstall, M.: SoC it to EM: ElectroMagnetic side-channel attacks on a complex System-on-Chip. In: Güneysu, T., Handschuh, H. (eds.) CHES 2015. LNCS, vol. 9293, pp. 620–640. Springer, Heidelberg (2015)

13. Google Inc.: Android Full Disk Encryption (2015). https://source.android.com/security/encryption/

14. Halderman, J.A., Schoen, S.D., Heninger, N., Clarkson, W., Paul, W., Calandrino, J.A., Feldman, A.J., Appelbaum, J., Felten, E.W.: Lest we remember: cold-boot attacks on encryption keys. Commun. ACM **52**(5), 91–98 (2009)

15. Hanley, N., Tunstall, M., Marnane, W.P.: Unknown plaintext template attacks. In: Youm, H.Y., Yung, M. (eds.) WISA 2009. LNCS, vol. 5932, pp. 148–162. Springer, Heidelberg (2009)

16. Ishai, Y., Sahai, A., Wagner, D.: Private circuits: securing hardware against probing attacks. In: Boneh, D. (ed.) CRYPTO 2003. LNCS, vol. 2729, pp. 463–481. Springer, Heidelberg (2003)

17. Jaffe, J.: A first-order DPA attack against AES in counter mode with unknown initial counter. In: Paillier, P., Verbauwhede, I. (eds.) CHES 2007. LNCS, vol. 4727, pp. 1–13. Springer, Heidelberg (2007)

18. Kaliski, B.: PKCS# 5: Password-based Cryptography Specification Version 2.0 (2000)

19. Kocher, P.C., Jaffe, J., Jun, B.: Differential power analysis. In: Wiener, M. (ed.) CRYPTO 1999. LNCS, vol. 1666, p. 388. Springer, Heidelberg (1999)

20. Linux Kernel Organization Inc.: Linux Kernel 4.3 Source Tree (2015). https://git.kernel.org/cgit/linux/kernel/git/torvalds/linux.git/log/?id=refs/tags/v4.3

21. Halcrow, M., Savagaonkar, U., Ts'o, T., Muslukhov, I.: Ext4 Encryption Design Document.   https://docs.google.com/document/d/1ft26lUQyuSpiu6VleP70_npaWdRfXFoNnB8JYnykNTg

22. Moradi, A., Poschmann, A., Ling, S., Paar, C., Wang, H.: Pushing the limits: a very compact and a threshold implementation of AES. In: Paterson, K.G. (ed.) EUROCRYPT 2011. LNCS, vol. 6632, pp. 69–88. Springer, Heidelberg (2011)

23. Percival, C.: Stronger Key Derivation via Sequential Memory-Hard Functions. Self-published, pp. 1–16 (2009)

24. Piret, G., Quisquater, J.-J.: A differential fault attack technique against SPN structures, with application to the AES and KHAZAD. In: Walter, C.D., Koç, Ç.K., Paar, C. (eds.) CHES 2003. LNCS, vol. 2779, pp. 77–88. Springer, Heidelberg (2003)

25. Rogaway, P.: Efficient instantiations of tweakable blockciphers and refinements to modes OCB and PMAC. In: Lee, P.J. (ed.) ASIACRYPT 2004. LNCS, vol. 3329, pp. 16–31. Springer, Heidelberg (2004)

26. Rogers, B., Chhabra, S., Prvulovic, M., Solihin, D.: Using address independent seed encryption and Bonsai Merkle trees to make secure processors OS- and performance-friendly. In: 40th Annual IEEE/ACM International Symposium on Microarchitecture, MICRO 2007, pp. 183–196, December 2007

27. Saarinen, M.-J.O.: Encrypted watermarks and Linux laptop security. In: Lim, C.H., Yung, M. (eds.) WISA 2004. LNCS, vol. 3325, pp. 27–38. Springer, Heidelberg (2005)

28. Saha, D., Mukhopadhyay, D., RoyChowdhury, D.: A diagonal fault attack on the advanced encryption standard. Cryptology ePrint Archive, report 2009/581 (2009). http://eprint.iacr.org/

29. Suh, G., Clarke, D., Gasend, B., van Dijk, M., Devadas, S.: Efficient memory integrity verification and encryption for secure processors. In: 36th Annual IEEE/ACM International Symposium on Microarchitecture, Proceedings 2003, MICRO-36, pp. 339–350, December 2003

Co-location Detection on the Cloud

Mehmet Sinan İnci[✉], Berk Gulmezoglu, Thomas Eisenbarth, and Berk Sunar

Worcester Polytechnic Institute, Worcester, MA, USA
{msinci,bgulmezoglu,teisenbarth,sunar}@wpi.edu

Abstract. In this work we focus on the problem of co-location as a first step of conducting Cross-VM attacks such as `Prime and Probe` or `Flush+Reload` in commercial clouds. We demonstrate and compare three co-location detection methods namely, cooperative Last-Level Cache (LLC) covert channel, software profiling on the LLC and memory bus locking. We conduct our experiments on three commercial clouds, Amazon EC2, Google Compute Engine and Microsoft Azure. Finally, we show that both cooperative and non-cooperative co-location to specific targets on cloud is still possible on major cloud services.

Keywords: Co-location on the cloud · Software profiling · Cache covert channel · Performance degradation attacks · Memory bus locking

1 Motivation

As the adoption of cloud computing continues to increase at a dizzying speed, so has the interest in cloud-specific security issues. A new security issue due to cloud computing is the potential impact of shared resources on security and privacy of information. An example is the use of caches to circumvent ASLR [11], one of the most common techniques to prevent control-flow hijacking attacks. Several other works target the exploitability of cryptography in co-located systems under increasingly generic assumptions. While early works such as [24] still required attacker and victim to co-reside on the same core within a processor, latest works [14,17] work across cores and managed even to drop the memory de-duplication requirement of `Flush+Reload` attacks [7,10,13,22]. Besides extracting cryptographic keys, there are plenty of other security issues explored in other related studies. Irazoqui et al. [16] study the potential of reviving the partially fixed Lucky 13 attack [8] by exploiting co-location.

All of the above attacks rely on the attacker's ability to co-locate with a potential victim. While co-location is an immediate consequence of the benefits of cloud computing (better utilization of resources, lower cost through shared infrastructure etc.), whether *exploitable* co-location is possible or easy has so far not been studied in detail. In his seminal work, Ristenpart et al. [18] studied the general feasibility of co-location in Amazon EC2, the most popular public cloud service provider (CSP) then and now, in detail. However, the cloud landscape has changed significantly since then: The EC2 has grown exponentially and operates data centers around the globe. A myriad of competitors have popped up,

© Springer International Publishing Switzerland 2016
F.-X. Standaert and E. Oswald (Eds.): COSADE 2016, LNCS 9689, pp. 19–34, 2016.
DOI: 10.1007/978-3-319-43283-0_2

all competing for the rapidly growing customer base [9]. CSPs are also more aware of the potential security vulnerabilities and have since worked on making their systems leak less information across VM boundaries. Furthermore, in their experiments, both co-located parties were colluding to achieve co-location. That is, both parties were willingly involved in communicating with the other to detect co-location. While being of high importance to show the feasibility in the first place, trying to co-locate with a specific and most likely unwilling target can be considerably harder. Since that initial work, until very recently only little work has dealt with a more detailed study on the difficulty of co-location. Therefore, we believe, the problem of co-location on cloud requires further in depth analysis examining different detection methods under diverse scenarios and access levels for the attacker.

1.1 Our Contribution

In this work we revisit the problem of co-location in public IaaS clouds. In particular we:

- study the co-location problem under two threat models in the Amazon EC2 Cloud, Google Cloud Engine and Microsoft Azure.
- develop a novel LLC software profiling tool that can detect an application or a library run by the non-cooperating co-located victim in the cloud, without the use of the memory de-duplication or any other memory sharing methods.
- demonstrate three co-location methods and compare their success rates on three popular public clouds.

2 Related Work

In the last few years several methods were proposed to detect co-location on commercial clouds [6,12,18,23,25]. These works use methods such as deducing co-location from instance and hypervisor IP address, hard disk drive performance degradation, network latency and L1 cache covert channel. However, in response to these works, most of the proposed techniques have been closed by public cloud administrators. Later Zhang et al. [23] were able to determine whether a particular user's VM had someone else co-residing in the same physical core. In particular, they utilized the well known `Prime and Probe` cache based side-channel technique to guess this information. However, the technique was applied in the upper level caches, thereby limiting its applicability to a physical core rather than the entire CPU or the machine. Furthermore, the technique was not tested in commercial clouds.

Shortly later, Bates et al. [6] demonstrated that a malicious VM can inject a watermark in the network flow of a potential victim. In fact, this watermark would then be able to broadcast co-residency information. Again, even though the technique proved to be extremely fast (less than 10 s), it was never tested in commercial clouds. Recently, Zhang et al. [25] demonstrated that Platform as a

Service (PaaS) clouds are also vulnerable to co-residency attacks. They used the `Flush+Reload` cache side-channel technique together with a non-deterministic finite automaton method to infer co-location with a particular server. The technique proved to be effective in commercial PaaS clouds like DotCloud or Open-Shift, but would never work in IaaS clouds where the memory de-duplication is not implemented, as in most of the commercial IaaS clouds.

Finally, İnci et al. [12] demonstrated that many of the previously utilized techniques in [18] are no longer exploitable. Nevertheless, they prove to detect co-location *across cores* in Amazon EC2 by monitoring the usage of the LLC with the `Prime and Probe` technique. To enable the co-location test, the authors make use of hugepages commonly available in commercial clouds. This feature provides a large memory space for the attacker to move and hit necessary addresses to prime cache sets. Also in 2015, Varadarajan et al. [20] investigated co-location detection in public clouds by triggering and detecting performance degradations of a web server using the memory bus locking mechanism. Simultaneously Xu et al. [21] used the same memory bus locking mechanism to explore co-location threat in Virtual Private Cloud (VPC) enabled cloud systems.

3   Threat Models

Here we briefly outline two attacks scenarios for cross-VM attacks on public clouds. The main difference between the two scenarios is whether the target is predetermined or not. As we shall see, this makes a significant difference in terms of the requirements and cost of a successful attack. We provide concrete examples for both scenarios.

Random Victim
In this scenario there are four steps:

1. **Co-location:** The attacker spins instances on the cloud until it is determined that the instance is not *alone*; i.e. is co-located with another VM. Here the goal is to maximize the probability and thereby reduce the cost of co-locating with a viable target. Cheaper instances that use fewer CPU cores tend to share the same hardware in greater numbers. Therefore these instances have a better chance of co-location with other customers. Since we do not discriminate between targets, this step is rather easy to achieve.

2. **Vulnerable Software Identification:** The attacker detects a software package in the co-resident VM vulnerable to cross-VM attacks by monitoring corresponding LLC sets of libraries, e.g. an unpatched version of a cryptographic library. Cache access/performance and more broadly fingerprinting based techniques do exist in the literature to make successful attacks in the cloud environment [15,19,25]. Here, instances with lower number of tenants are less noisy therefore have higher success rate of library detection and the actual attack.

3. **Cross-VM Secret Extraction:** Here the attacker runs one of the cross-VM attacks [12,14] on the identified target. By exploiting cross-VM leakage

the attacker would be able to recover a sensitive information ranging from specialized pieces of information such as cryptographic keys, to higher level information such as browsing patterns, shopping cart, system load or any sensitive information of value. Noise plays a significant role in reliability of the extraction technique. Since co-location (first step) is easy to achieve, it is (almost) always advisable to opt for a less populated low noise instance to improve the chance of a successful attack in the later steps.

4. **Value Extraction:** The result is some sensitive information that can be turned into value with additional mild effort. For example, some information is valuable in its own right and can be converted into money with little or no effort, e.g., bitcoins, credit card information, credentials for online banking. Some others require further effort such as TLS session encryption key (secret key), e.g. for a Netflix streaming session. If the recovered secret is a private key of a public key encryption scheme (e.g. RSA secret key used a TLS handshake) the attacker needs the identity of the owner (website/company) to have further use for the secret key. In this case he may check the private key against public key repositories for noise correction and target identification.

Targeted Victim

This is the complementary scenario where we are given some identification information about the target.

1. **IP Extraction:** The attacker wants to focus its cycles on a server or a group servers that belong to an individual, cloud backed business, e.g. Dropbox or Netflix, or group/entity, e.g. dissidents of a political party. Here we assume that the attacker is capable of resolving the identification information to an IP or group of IPs of the target. In practice, this can be achieved rather easily by using public information and by using simple commonly available network tools such as `traceroute/tracepath`, `nmap` etc.

2. **Targeted Co-location:** The attacker creates instances on the cloud until one is co-located with the target instance on the same physical machine. The identification information of the victim, e.g. IP address, is used for co-location detection. For instance, using the IP the attacker can query the server creating CPU load and then run co-location tests. While co-location detection will be easier in this scenario due to the trigger; we will need many more trials to land on the same physical machine as the victim[1]. Nevertheless, we can accelerate targeted co-location by *searching*, for instance, only in the same region as the victim instance using the publicly available AWS IP lists [1]. Further, we can obtain finer grain information about the target's location simply by running `traceroute` or `tracepath` on the victim IP.

3. **Vulnerable Software Identification:** Since we know the identity of our target, it is safe to assume that we have some rudimentary understanding of

---

[1] Note that if the physical machine is already filled with the maximum number of allowed instances, then co-location may not be possible at all. In this case a clever albeit costly strategy would be to first mount a denial of service attack causing the target instance to be replicated and then try co-locating with the replicas.

the victim's setup including OS, communication and security protocols used etc. Even if this is not the case, it would be possible to run a discovery stage to survey the victim machine using its IP and by detecting process fingerprints through cross-VM leakage.

4. **Value Extraction:** The attacker exploits cross-VM leakage to recover sensitive information. Further processing may allow to enhance quality of the recovered data using publicly available information. For instance, a noisy private key can be processed with the aid of the public key contained in the certificate belonging to the target to remove any imperfections.

4   Overview: Co-location Detection Methods

4.1   LLC Covert Channel

The LLC is shared across all cores in most modern CPUs and is semi-transparent to all VMs running on the same machine. By semi-transparent, we mean that all VMs can utilize the entire LLC but cannot read each other's data. We exploit this behavior to establish a covert channel between VMs in cloud. The covert channel works by two VMs writing to a specific set-slice pair in the LLC and detecting each other's accesses. LLC set address can easily be deduced from the virtual addresses available to VMs using hugepages as done in [12,14,17]. The cache slice on the other hand, cannot be determined with certainty unless the slice selection algorithm of the CPU is known. However, the covert channel can still work by priming more sets and accessing lines that go to the targeted set, regardless of its slice.

Prime and Probe: In the LLC, the number of lines required to fill a set is equal to the LLC associativity. However, when multiple users access the same set, one will notice that fewer than 20 lines are needed to observe evictions. By running the following test concurrently on multiple instances, we can verify co-location. The test works as follows:

- Calculate the set number by using the address bits that are not affected by the virtual to physical address translation. Prime a memory block M_0 in the set.
- Access more memory blocks $M_1, M_2, \ldots, M_n$ that go to the same set. Note that since the slice selection algorithm for the specific CPU is necessary to address a set/slice pair with certainty, the number of memory blocks n needs to be larger than the set associativity times the number of slices.
- Access the memory block M_0 and check for eviction from the LLC. If evicted, we know that the required b memory blocks that fill the set are among the accessed memory blocks $M_1, M_2, \ldots, M_n$.
- Starting from the last memory block accessed, remove one block and repeat the above protocol. If M_0 still has high access time, M_i does not reside in the same slice. If b_0 is now located in the cache, we know that b_i resides in the same cache slice as b_0 and therefore go to the same set.

– Once the b memory blocks that fill a slice are identified, we just access additional memory blocks and check whether one of the primed b memory blocks has been evicted, indicating that they collide in the same slice.

The covert channel works by continuously accessing data that goes to a specific cache set and measuring the access time to determine if a newly accessed data has evicted an older entry from the set. Due to this continuous cache line creation, when the second party makes accesses to the monitored set, they are detected. In general, if there is no noise present, the number of lines that can go to a set without triggering an eviction is equal to the associativity of the cache, assuming a first-in first-out (FIFO) cache replacement policy is employed.

When two VMs try to fill the same set, they have to access less number of data blocks to fill the specified cache hence detecting the co-location. Using the number of blocks necessary to fill a specific set with and without another instance interfering, we calculate a co-location confidence ratio.

4.2 Software Profiling on LLC

The software profiling method works in a realistic setting with minimal assumptions. The method works in a non-cooperative scenario where the target does not participate in a covert communication and continues its regular operation. The method does not require memory de-duplication or any form of shared libraries. It employs the `Prime and Probe` to monitor and profile a portion of the LLC while a targeted software is running. As for the memory addressing, we profile the targeted code address as a relative address to the page boundary. Since the targeted library will be page aligned, target code's relative address (the page offset) will remain the same between runs. Using this information, we can reduce our search space in the detection stage. Therefore, we need to monitor only 320 different set-slice pairs such as $X \mod 64 = Y$ where X is 320 different set numbers (since we have 10 cores and 32 different set numbers satisfying the equation) and Y is the first 6 bits (the first 6 bits of the LLC set number is directly converted to physical address) of the set number for the desired function.

For the RSA detection, the slice-selection algorithm of the CPU is required to locate the targeted multiplication code in the LLC in a reasonable time. Without the algorithm, it would take too much time to monitor potential cache sets. For our experiments, we have used the algorithm that was reverse engineered by İnci et al. in [12].

In summary, there are two stages to the software profiling on LLC;

– **Profiling Stage:** The first step of the profiling is to monitor the targeted LLC sets while the profiled code, the software is not running. The purpose of this stage is to measure the idle access time of 20 lines for each set to have a threshold to detect whether there is a cache miss or not in the next stage.
– **Detection Stage:** We send RSA decryption requests to candidate IPs in order to discover the IP address of the victim. After triggering the decryption we

begin to monitor the portion of LLC to detect accesses due to the decryption. If we detect accesses in targeted set-slice pairs then we know that the correct IP address is found. As a double check, in addition to the RSA detection, we also detect AES encryption. In order to so we monitor another portion of the LLC where the AES T-tables potentially reside. And if the victim is co-located with the attacker, we can detect and monitor these T-table accesses.

4.3 Memory Bus Locking

The memory bus locking method exploits atomic instructions therefore we explain these special instructions shortly in the following.

Atomic Operations: Atomic operations are defined as indivisible, uninterrupted operations that appear to the rest of the system as instant. When operating directly on memory or cache, an atomic operation prevents any other processor or I/O device from reading or writing to the operated address. This isolation ensures computational correctness and prevents data races. While all instructions on single thread systems are automatically atomic, there is no guarantee of atomicity for regular instructions in multi-thread systems as used in almost all modern systems. In these systems, an instruction can be interrupted or postponed in favor of another task. The rescheduling, interruption and operating on the same data can cause pipeline and cache coherency hazards. Therefore the atomic operations are especially useful on multi-thread systems and parallel processing.

In older x86 systems, processor locks the memory bus completely until the atomic operation finishes, whether the data resides in the cache or in the memory. While ensuring atomicity, the process results in a significant performance hit. In newer systems - prior to Intel Nehalem and AMD K8 - memory bus locking was modified to reduce this penalty. In these systems, if the data resides in cache, only the cache line that holds the data is locked. This lock results in a very insignificant system overhead compared to the performance penalty of memory bus locking. However, when the operated data surpasses cache line boundary and resides in two cache lines, more than a single cache line has to be locked. In order to do so, memory bus locking is again employed. After Intel Nehalem and AMD K8, shared memory bus was replaced with multiple buses with non-uniform memory access bridge between them. While getting rid of the memory bottleneck for multiprocessor systems, this also invalidated the memory bus locking. Now, when a multi-line atomic cache operation has to be performed, all CPUs has to coordinate and flush their ongoing memory transactions. This emulation of memory bus locking results in a significant performance hit.

In x86 architecture, there are many instructions that can be executed atomically with a lock prefix are ADC, ADD, AND, BTC, BTR, BTS, CMPXCHG, DEC, FADDL, INC, NEG, NOT, OR, SBB, SUB, XADD, XOR. Also, XCHG instruction executes atomically when operating on a memory location, regardless of the LOCK use. In order to maximize the flushing penalty, we tested all atomic instructions available to the platforms and measured how long each instruction takes to execute. Since the flushing is succeeded with the atomic operation itself, longer the instruction executes, stronger the performance hit becomes. Therefore we have used the XADDL instruction that resulted in the strongest penalty. In short, we employ this mechanism to slow down a server process running in the cloud and detect co-location **without cooperation** from the victim side.

Cache Line Profiling Stage: Our attack is CPU-agnostic and employs a short, preliminary cache profiling stage. This stage eliminates the need for the information like the cache line size and the cache access time. Our purpose here is to obtain data addresses that span multiple cache lines hence triggers a bus lock. First, we allocate a block of small, page-aligned memory using *malloc*. After the allocation, we start performing atomic operations on this block in a loop of 256 since no modern cache line is expected to be larger than 256 bytes. In each loop, we move our access pointer by one and record atomic operation execution times. When we observe a time larger than the pre-calculated average, we record the address. After all 256 addresses are tested, we obtain a list of addresses that span across multiple cache lines. Later during the locking stage, we operate only on these addresses rather than a continuous array, making the attack more efficient.

Dual Socket Problem: Memory bus locking works on systems with multiple CPU sockets. Even further, our tests reveal that the bus locking penalty clearly reveals whether the target and the attacker run in the same socket or not. As seen in Fig. 1, the memory access time is clearly distinguishable between same socket and different socket locks. On a dual socket system with two Intel Xeon E5-2609 v2 CPUs with 2 cores each. Note that this information is significant to the attacker since an architectural attack using the LLC requires the attacker and the target to be running in the same socket.

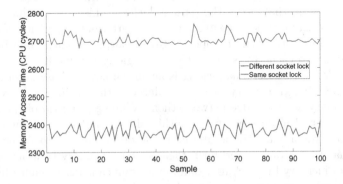

Fig. 1. The memory access times during a bus lock triggered with the XADDL instruction. Red and blue lines respectively represent access times when the attacker resides in the same socket (different core) and different sockets. (Color figure online)

5   Experimental Approach and Results

5.1   Co-location Results in Commercial Clouds

In all three aforementioned commercial clouds, we have launched 4 accounts with 20 instances per account, achieving co-location in each cloud. Also note that, we only classify the instances running in the same CPU socket as co-located and ignore the ones running on different sockets.

Amazon EC2: In Amazon EC2 we used **m3.medium** instance types that have balanced CPU, memory and network performance. This instance type holds 1 vCPU, 3.75 GB of RAM and 4 GB of SSD storage. According to Amazon EC2 Instance Types web page [4], these instances use 10 core Intel Xeon E5-2670 v2 (Ivy Bridge) processors.

Out of 80 instances launched, we have obtained 7 co-located pairs and one triplet verified by the tests. Moreover, we have tried to co-locate with instances that have launched previously. Surprisingly, we have been able to co-locate with instances that have launched 6 months prior.

Google Compute Engine: In GCE, we used **n1-standard-1** type instances running on 2.6 GHz Intel Xeon E5 (Sandy Bridge), 2.5 GHz Intel Xeon E5 v2 (Ivy Bridge), or 2.3 GHz Intel Xeon E5 v3 (Haswell) processors according to [5]. Out of 80 instances launched, we have obtained only 4 co-located pairs.

Microsoft Azure: In Azure, we used **extra small A0** instance types with 1 virtual core, 750 MB RAM, maximum 500 IOPS and 20 GB disk storage that is not specified as neither SSD nor HDD [2]. Out of 80 instances launched, we have obtained only 4 instances that were co-located. However, this was partly due to the highly heterogeneous CPU pool that Azure employs. Our first account had instances with AMD Opteron CPUs while the second had Intel E5-2660 v1 and the last two had Intel E5-2673 v3. Naturally, we could only achieve co-location among instances that have the same CPU model. Out of 40 Intel E5-2673 v3 instances, we detected 4 co-located instances.

5.2   LLC Covert Channel

In the following, we present the results in GCE. The confidence ratio is highest at 1 as seen in Fig. 2. There are 8 instances (meaning 4 pairs) that have higher than 50 % confidence ratio among 80 and the co-located pairs are found by binary search at the end. Hence, it is confirmed that they are indeed co-located with each other.

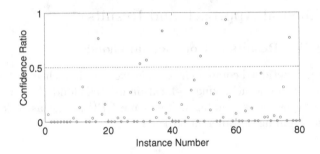

Fig. 2. GCE LLC Test Confidence Ratio Comparison

5.3   LLC Software Profiling

We conducted the LLC Software Profiling experiments on the co-located Amazon EC2 instances with 10 core E5-2670 v2 processors. As for the software target, in order to demonstrate the versatility of the attack, we chose the RSA (Libgcrypt version 1.6.2) that uses sliding window exponentiation and the AES (OpenSSL version 1.0.1g, C implementation) that uses T-tables. Note that the detection method is not limited to these targets since the attacker can run and profile any software which uses shared library in his instance and perform the attack.

For the RSA detection, the slice-selection algorithm of the CPU is required to locate the targeted multiplication code in the LLC within reasonable time. In our experiments, we have used the algorithm that was reverse engineered by İnci et al. in [12]. The first step of the profiling is to monitor the targeted LLC sets while the profiled code, RSA is not running. After the regular operation of sets are observed, the RSA request is sent to several IP addresses, starting from attacker's own subnet. As soon as the request is sent, the profiling starts and traces are recorded by the **Prime and Probe**. If the RSA decryption is running on the other VM, the pattern of multiplication can be observed as in Fig. 3. In general, the multiplication is performed between 2000–8000 traces. In these traces, we look for the delta of two profiles for each set-slice pair. In Fig. 4, the difference between two profiles is illustrated for two co-located instances. Both figures show that there are two set-slice pairs with significantly higher access times (4–8 cycles) in average of 10 experiments. Hence, it can be concluded that these two sets are used by RSA decryption and this candidate instance is probably co-located with the attacker.

After we obtain IP addresses of several co-location candidates, we trigger AES encryption by sending random ciphertexts and at the same time monitor the LLC. For this part of the detection stage, since AES encryption is much faster than RSA decryption we can only catch one access to monitored T-table position. Hence, we send 100 AES encryption requests to each instance in the IP list. If we observe 90 % cache miss for one of the set-slice pairs, it can be concluded that the AES encryption is performed by the co-located instance, as seen in Fig. 3(b).

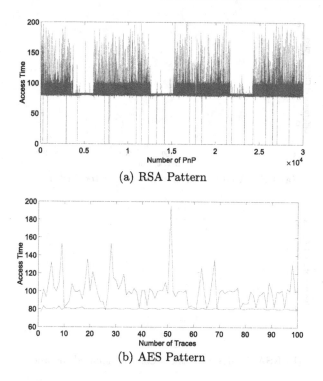

(a) RSA Pattern

(b) AES Pattern

Fig. 3. Red and blue lines represent idle and RSA decryption/AES encryption access times respectively (Color figure online)

5.4 Memory Bus Locking

The performance degradation due to the memory bus locking is application specific. Therefore we tested various applications as seen in Table 1 to see how each one is affected. As expected, the applications with frequent memory accesses are more affected by the locking. For example, the GnuPG which mostly uses the ALU and does seldom memory accesses slowed down only by 29 %. An Apache web server that frequently loads content from memory on the other hand has a slowdown by the factor of 4.28.

In addition to specific software performance degradation, we also measured the effect of multiple locks executed in parallel. To do so, we have used the openmp parallel programming API [3] and ran the lock in multiple threads. Figure 5(d) shows the memory access times when 0 to 8 locks run in parallel. As the figure shows, the first lock does slowdown the memory accesses by 100 % while the second and third locks do not further degrade the memory performance. However, after fourth and fifth locks, we observe an even stronger degradations.

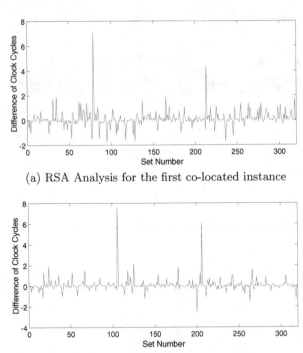

(a) RSA Analysis for the first co-located instance

(b) RSA Analysis for the second co-located instance

Fig. 4. The difference of clock cycles between base and RSA decryption profiling for each set-slice pairs over 10 experiments

Table 1. Application slowdown on an Intel Xeon 2640 v3 due to memory bus locking triggered on a single core.

Process	Normalized execution time
Apache	4.28×
PHP	0.1×
GnuPG	0.29×
HTTPerf	0.29×
Memory access	5.38×
RAMSpeed int	5.01×
RAMSpeed fp	4.88×
Media stream	2.36×

5.5   Comparison of Detection Methods

As explained in Sect. 3, co-location can be exploited in both random and targeted victim scenarios. Malicious Eve can directly look for attack vectors to steal information from her neighbors or she can go after a specific target and spin up

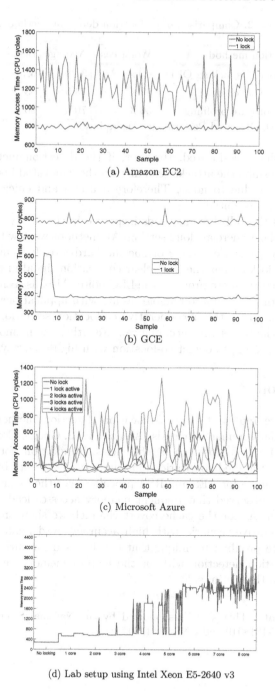

(a) Amazon EC2

(b) GCE

(c) Microsoft Azure

(d) Lab setup using Intel Xeon E5-2640 v3

Fig. 5. Memory access times with and without an active memory bus lock of (a) Amazon EC2 m3.medium instance (b) GCE n1-standard1 instance (c) Microsoft Azure A0 instance (d) Lab setup (Intel E5-2640 v3) (Color figure online)

Table 2. Comparison of co-location detection methods.

Detection method	Worst case	Average	Best case
Memory bus locking OPD[a]	0.1×	3.28×	6.1×
LLC covert channel	53%	73.5%	93%
LLC software profiling	50%	70%	90%

[a]OPD: Observed Performance Degradation

instances until she is co-located. However, if the detection method does not provide reliable results, the attacker can discard the co-located instances or even have false positives due to noise. Therefore a useful and efficient co-location detection method is essential.

Table 2 shows that all three methods inspected in this study work with high accuracy in a real commercial cloud setting. All methods work with minimalistic requirements, no hypervisor access or specific hardware. In comparison, while the memory bus locking has the least clear co-location signal in the worst case, other two methods are more prone to the LLC noise. Also, as seen in Table 1 the memory bus locking gives more reliable results with applications with frequent memory accesses. So for the uncooperative co-location scenario, depending on the workload of the target instance, one can use either the memory bus locking or the software profiling to detect co-location with high accuracy.

6   Conclusion

In conclusion, we represent three co-location detection methods working in three most popular commercial clouds (Amazon EC2, Google Compute Engine, Microsoft Azure) and compare their efficiencies. In addition, for the first time we have achieved targeted co-locations in Amazon EC2 Cloud by applying the LLC software profiling for AES and RSA processes. For the memory bus locking method, we have observed that frequent memory accesses lead to more significant degradation. As for the cache covert channel, we show that the method works in a cooperative scenario with high accuracy. And finally we presented the LLC software profiling technique that can be used for variety of purposes including co-location detection without the help of memory de-duplication or cooperation from the victim side.

Acknowledgments. This work is supported by the National Science Foundation, under grants CNS-1318919 and CNS-1314770.

References

1. AWS   IP   Address   Ranges.   http://docs.aws.amazon.com/general/latest/gr/aws-ip-ranges.html
2. Microsoft Azure Sizes for virtual machines. https://azure.microsoft.com/en-us/documentation/articles/virtual-machines-size-specs/

3. The OpenMP API specification for parallel programming
4. Amazon EC2 Instances (2016). http://aws.amazon.com/ec2/instance-types/
5. Google Compute Engine Instance Types (2016). https://cloud.google.com/compute/docs/machine-types
6. Bates, A., Mood, B., Pletcher, J., Pruse, H., Valafar, M., Butler, K.: On detecting co-resident cloud instances using network flow watermarking techniques. Int. J. Inf. Secur. **13**(2), 171–189 (2014). http://dx.doi.org/10.1007/s10207-013-0210-0
7. Benger, N., van de Pol, J., Smart, N.P., Yarom, Y.: "Ooh Aah.. Just a Little Bit": a small amount of side channel can go a long way. In: Batina, L., Robshaw, M. (eds.) CHES 2014. LNCS, vol. 8731, pp. 75–92. Springer, Heidelberg (2014)
8. Fardan, N.J.A., Paterson, K.G.: Lucky thirteen: breaking the TLS and DTLS record protocols. In: Security and Privacy, pp. 526–540 (2013)
9. Gaudin, S.: Public cloud market ready for 'hypergrowth' period. Computerworld Article, April 2014. http://www.computerworld.com/article/2488572/cloud-computing/public-cloud-market-ready-for-hypergrowth-period.html
10. Gülmezoglu, B., İnci, M.S., Apecechea, G.I., Eisenbarth, T., Sunar, B.: A faster and more realistic flush+reload attack on AES. In: COSADE, pp. 111–126 (2015)
11. Hund, R., Willems, C., Holz, T.: Practical timing side channel attacks against kernel space ASLR. In: Proceedings of the 2013 IEEE Symposium on Security and Privacy, pp. 191–205 (2013). http://dx.doi.org/10.1109/SP.2013.23
12. İnci, M.S., Gulmezoglu, B., Irazoqui, G., Eisenbarth, T., Sunar, B.: Seriously, get off my cloud! Cross-VM RSA key recovery in a public cloud. Technical report. http://eprint.iacr.org/
13. Irazoqui, G., İnci, M.S., Eisenbarth, T., Sunar, B.: Fine grain Cross-VM attacks on Xen and VMware. In: 2014 IEEE Fourth International Conference on Big Data and Cloud Computing (BdCloud), pp. 737–744, December 2014
14. Irazoqui, G., Eisenbarth, T., Sunar, B.: S$A: a shared cache attack that works across cores and defies VM sandboxing? And its application to AES. In: IEEE S&P (2015)
15. Irazoqui, G., İnci, M.S., Eisenbarth, T., Sunar, B.: Know thy neighbor: crypto library detection in cloud. In: Proceedings on Privacy Enhancing Technologies, vol. 1, no. 1, pp. 25–40 (2015)
16. Irazoqui, G., Inci, M.S., Eisenbarth, T., Sunar, B.: Lucky 13 Strikes Back. In: ASIA CCS 2015, pp. 85–96 (2015)
17. Liu, F., Yarom, Y., Ge, Q., Heiser, G., Lee, R.B.: Last-level cache side-channel attacks are practical. In: IEEE S&P, pp. 605–622 (2015)
18. Ristenpart, T., Tromer, E., Shacham, H., Savage, S.: Hey, you, get off of my cloud: exploring information leakage in third-party compute clouds. In: CCS 2009, pp. 199–212 (2009)
19. Suzaki, K., Iijima, K., Yagi, T., Artho, C.: Memory deduplication as a threat to the guest OS. In: Proceedings of the Fourth European Workshop on System Security, p. 1. ACM (2011)
20. Varadarajan, V., Zhang, Y., Ristenpart, T., Swift, M.: A placement vulnerability study in multi-tenant public clouds. In: 24th USENIX Security Symposium, USENIX Security 2015, Washington, D.C., pp. 913–928 (2015)
21. Xu, Z., Wang, H., Wu, Z.: A measurement study on co-residence threat inside the cloud. In: 24th USENIX Security, pp. 929–944 (2015)
22. Yarom, Y., Falkner, K.: FLUSH+RELOAD: a high resolution, low noise, L3 cache side-channel attack. In: USENIX Security 2014, pp. 719–732 (2014)
23. Zhang, Y., Juels, A., Oprea, A., Reiter, M.K.: HomeAlone: co-residency detection in the cloud via side-channel analysis. In: IEEE S&P (2011)

24. Zhang, Y., Juels, A., Reiter, M.K., Ristenpart, T.: Cross-VM side channels and their use to extract private keys. In: CCS 2012, pp. 305–316 (2012)
25. Zhang, Y., Juels, A., Reiter, M.K., Ristenpart, T.: Cross-tenant side-channel attacks in PaaS clouds. In: CCS, pp. 990–1003 (2014)

Simple Photonic Emission Attack
with Reduced Data Complexity

Elad Carmon[1], Jean-Pierre Seifert[2,3], and Avishai Wool[4(✉)]

[1] Tel-Aviv University, 69978 Tel-Aviv, Israel
eladca@gmail.com
[2] Security in Telecommunications, Technische Universität Berlin, Berlin, Germany
Jean-Pierre.Seifert@telekom.de
[3] FhG SIT, Darmstadt, Germany
[4] Tel-Aviv University, 69978 Tel-Aviv, Israel
yash@eng.tau.ac.il

Abstract. This work proposes substantial algorithmic enhancements to the SPEA of Schlösser et al. [15] by adding cryptographic post-processing, and improved signal processing to the photonic measurement phase. Our improved approach provides three crucial benefits: (1) For some SBox/SRAM configurations the original SPEA method is unable to identify a unique key, and terminates with up to 2^{48} key candidates; using our new solver we are able to find the correct key regardless of the respective SBox/SRAM configuration. (2) Our methods reduce the number of required (complex photonic) measurements by an order of magnitude, thereby shortening the duration of the attack significantly. (3) Due to the unavailability of the attack equipment of Schlösser et al. [15] we additionally developed a novel Photonic Emission Simulator which we matched against the real equipment of the original SPEA work. With this simulator we were able to verify our enhanced SPEA by a full AES recovery which uses only a small number of photonic measurements.

1 Introduction

1.1 Background

While the phenomena of photonic emission from switching transistors in silicon is actually a very old one, cf. [5,12], the role of photons in cryptography as a practical side channel source has just recently emerged as a novel research direction, cf. [3,9,10,15,16]. Thus, it is important to include photonic side channels in future hardware evaluations of security ICs.

However, so far only the first steps within this direction have been successfully achieved: The work of [3,9,10,15,16], showed that the required equipment to carry out successful SPEA or DPEA against real world ICs is comparable in price to that of normal Power Analysis equipment.

A. Wool—Supported in part by a grant from the Interdisciplinary Cyber-Research Center at Tel Aviv University.

© Springer International Publishing Switzerland 2016
F.-X. Standaert and E. Oswald (Eds.): COSADE 2016, LNCS 9689, pp. 35–51, 2016.
DOI: 10.1007/978-3-319-43283-0_3

This is where the current paper fits in and continues the current state of the art in a better understanding of the Photonic Side Channel. It takes the next step by precisely characterizing a very low number of selected plaintexts as required for the respective photonic measurements and also relating the resulting measurements in terms of their SNR to the eventual workload of the final cryptographic key reconstruction phase.

1.2   Related Work

Photonic emission in silicon is a known physical phenomena which has been studied since the 1950s [12]. Specifically in the failure analysis community, hot-carrier luminescence has primarily been used to characterize implementation and manufacturing faults and defects [7,17]. Here, the technologies of choice to perform backside analysis are PICA (Picosecond Imaging Circuit Analysis) [1] and SSPDs (Superconducting Single Photon Detectors) [18]. Both technologies are able to capture photonic emissions with high performance in their respective field, but carry the downside of immense cost and complexity.

One of the first uses of photonic emissions in CMOS in a cryptographic application was presented in 2008 [8]. However, the authors increased the voltage supply to 7 V operating voltage, which is above the chips maximum limit for voltage. The authors utilize PICA to spatially recover information about binary additions related to the AddRoundKey operation of AES running on a 0.8 μm PIC16F84A microcontroller. As the authors state, such a PICA device "is available in several laboratories, for example, in the French space agency CNES". Employing PICA in this manner led to enormous acquisition times. This is especially true considering the size of the executed code. It took the authors 12 h to recover a *single* potential key byte [8]. In 2011, an integrated PICA system and laser stimulation techniques were used to attack a DES implementation on an FPGA [6]. The authors proved that the optical side channel might be used for differential analysis. However, the analysis strongly relied on a specific implementation of DES in which registers were always zeroed before their use. The results required a differential analysis and a full key recovery was also not presented. As the authors note, the use of equipment valued at more than 2,000,000 Euros does not make such an analysis particularly relevant.

Nevertheless, recently, a real breakthrough was achieved by [15,16]. This work presented a novel low-cost optoelectronic setup for time - and spatially resolved analysis of photonic emissions. The authors also introduced a corresponding methodology, named Simple Photonic Emission Analysis. They successfully performed such analysis of an AES implementation and were able to recover AES-128 keys by monitoring memory accesses. This work was also extended to AES-192 and AES-256 [16]. The same research group also introduced Differential Photonic Emission Analysis and presented a respective attack against AES-128 [10]. They successfully revealed the entire secret key with their DPEA. In 2015 Bertoni et al. [3] offered an improved Simple Photonic Emission Analysis, monitoring a different section of the SRAM logic. However, they assumed a specific SRAM structure which contains only single byte in every row. Their simulations do not model the physical environment but rather an ideal model in which the value of every bit can be identified. They also described an attack of masked AES, however the attack is unrealistic since it assumes monitoring the photonic emission of a single experiment.

A side channel analysis using memory access patterns is reminiscent of the field of cache attacks. For instance, the first "real world" cache-based chosen plaintext attacks on AES were carried on OpenSSL implementations [2,13].

1.3 Contributions

In this work we enhance the original SPEA of Schlösser et al. [15] by adding cryptographic post-processing and an improved signal processing to the measurements phase. We call the resulting attack Enhanced SPEA, or E-SPEA for short.

Our first contribution is to record the photonic side-channel leaks from the first *two* AES rounds, covering 32 SBox activations. We show that these leakages embed enough constraints to allow the identification of the complete key, regardless of the placement of the SBox array in SRAM. This is in contrast to the original SPEA, which terminates with up to 2^{48} key candidates for certain SRAM configurations. Furthermore, taking advantage of the slow diffusion properties in the first AES round, we are able to mount this attack very efficiently, with a time complexity of 2^{20}. Our optimized cryptographic solver finds the correct key within minutes on a standard PC.

Next, we devise a strategy for choosing optimal plaintexts, that causes the photonic side-channel to produce constraints (specific SRAM accesses) which enable our solver to work very quickly for all SRAM configurations. We collect the necessary constraints with only 32 plaintexts, instead of the 256 plaintexts required by Schlösser et al. [15].

Moreover, we developed a special signal-processing decoder that automatically calibrates certain internal thresholds—relying on our chosen plaintext strategy. The decoder works even when the SNR is low, adjusting its thresholds differently to match the requirements of the cryptographic solver. To do so, the decoder uses a different (auto-calibrated) threshold for each AES round. Using the combination of our carefully crafted decoder and solver, we can trade off the number of measurements against the solver's running time: fewer measurements (i.e., a lower SNR) cause a longer running time—but without missing the correct key.

The combination of the above contributions provides two main benefits.

1. We are always able to quickly find the correct key, regardless of the SRAM configuration.
2. Our methods reduce the number of required *optical* measurements dramatically by an order of magnitude, and thus we are able to shorten the duration of the attack significantly.

Also, in order to validate our attacks we built a Monte-Carlo simulator of the underlying physics of the photonic emissions, with a noise model which incorporates

- internal noise within the detector,
- external noise from nearby transistors, and
- other effects.

We validated our simulator against the results as reported in Schlösser [14]. Our simulator can be used to explore alternative lab setups, taking into account various critical parameters such as the lens area, height above the chip, supply voltage, ambient temperature, and equipment sensitivity.

We also believe that our photonic emission simulator is of independent interest and is of great value for the research community lacking (so far) the optical equipment as described within Schlösser [14].

Organization. The organization of the present paper is as follows. Section 2 introduces the SPEA on AES. Section 3 describes our cryptographic solver. Section 4 explains our choice of plaintexts. Section 5 describes the Auto-calibrating decoder. Section 6 describes our performance evaluation, and we conclude in Sect. 7. The description of the photonic emission simulator can be found in the full technical report [4].

2   The Photonic Side Channel in AES

2.1   The SRAM and Its Use in AES

SRAM is a common type of volatile memory found in many ICs. The SRAM is built from memory cells arranged in rows and columns, and every memory cell can be approached using a row/column access logic. In particular, the access logic for each SRAM row includes a so called row-access transistor, which is activated whenever the IC needs to access any cell in that SRAM row. Due to to this functionality, i.e., enabling an entire row, the respective row-access transistor is very strong. This means that the photonic emission of this transistor is by magnitudes larger than the individual SRAM cells by itself. For a thorough introduction into SRAM and its physical implementation details we refer the reader to [19].

The number of bytes in an SRAM row depends on the underlying SRAM architecture. In [15] the authors found that on an AT-Mega328P a single SRAM row consists of 8 bytes, whereas an ATXMega128A1 stores 16 bytes in an entire row. Figure 1(a) shows a photo of the SRAM, with a row width of 8 bytes.

A central component of the AES cipher is the SBox. This is an array of 256 bytes which is most often implemented as a lookup table. In each AES round the algorithm performs 16 SBox lookups. In many ICs implementing AES in software the entire SBox array is placed in SRAM.

In this paper we will denote the SRAM row width by ω. In general the SBox starts at an *offset* within an SRAM row, $0 \leq$ offset $\leq \omega - 1$, and occupies

$L = \lceil 256/\omega \rceil$ rows (see Fig. 1(b)). When $\omega = 8$, depending on the offset, we have $L = 32$ or $L = 33$. As we shall see, the value of the offset has an impact on the SPEA.

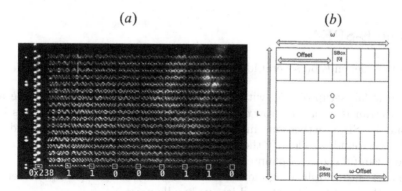

Fig. 1. The SRAM memory in (a) captured with a CCD by the courtesy of [15]. The row-access transistors appear to the left of the SRAM cells. In (b), a schematic of the SRAM section containing the SBox in L rows, ω cells per row and starting at some offset value.

2.2 Simple Photonic Emission Analysis (SPEA)

Monitoring the access patterns to the SRAM rows allows the SPEA as presented in [15]. Towards this goal, [15] first used a simple CCD camera approach to initially map the respective IC's layout, locating the SRAM memory, and specifically, the memory rows containing the SBox array and the offset value, cf. [11]. Hereafter, they placed a NIR (Near Infra Red) photon detector offering time resolved measurements over the row access transistor of some SRAM row containing SBox values. We call the SBox row on which the detector is placed *the detectable row*, and denote its number by d $(1 \le d \le L)$. The authors ran the AES algorithm M times (by actually resetting the IC M times), encrypting the same plaintext. Consider one of the 16 SBox activations of the first AES round for plaintext byte p_i and key byte k_i. If the detector identifies an activation for SBox$(p_i \oplus k_i)$, then there are ω options for $p_i \oplus k_i$ and since the plaintext is known, they have ω options for k_i.

Using all possible plaintext bytes $\{0, 1, \ldots, 255\}$ (M times each) they revealed sets of ω potential candidates for every byte of the key, then they analyzed each key byte separately, intersecting sets of candidates for every key byte reducing the number of potential candidates. The success of the SPEA method depends on two factors:

1. Using a large enough number of measurements M, providing a sufficient SNR.

2. The offset value. The SPEA works best when the offset is odd. In other cases its performance is limited, and in particular when offset = 0 the number of candidates for every key byte can't be reduced below ω candidates for each byte, resulting in ω^{16} key candidates.

3   The E-SPEA

Our attack depends on several ideas:

1. Use the lab setup of [15], with a NIR photon detector placed over the row access transistor of some row d in the SBox, to record the photonic emissions from the SBox activations in *2 full AES rounds* and use the dependence between rounds to identify the correct key.
2. Use a careful choice of plaintexts to quickly reduce the entropy.
3. A novel auto-thresholding method, based on the choice of plaintexts, lets us avoid the need to calibrate and lets us handle noise.

During the AES encryption process, there are ten rounds, each accessing SRAM memory to use the SBox array. In every round 16 bytes of the current state matrix are replaced by 16 bytes copied from the SRAM memory using the SBox as a lookup table.

Following [15] we place a detector over the location of the transistor controlling access to a row of SRAM containing ω cells of the SBox array. Thus each of the 16 SBox accesses per AES round has a $\approx 1/L$ probability that the row on which the detector is located ("the detectable row") will be accessed, assuming a random plaintext. Our attack requires knowing the offset value (recall Fig. 1(a)) and the row number (d) of the detectable row.

3.1   The Attack Structure

The attack activates the AES IC to encrypt plaintexts of the form $\{a, a, \ldots, a\}$ (all plaintext bytes are the same) for different values of a. For each key byte k_j, if the detectable row is accessed in the first AES round while looking up state byte j in the SBox, we obtain a constraint on the possible value of k_j, which reduces the number of possibilities for its value from 256 to ω. In [15] the authors iterated over all 256 plaintext options, guaranteeing that the detectable row is accessed at least once for every key byte in the first AES round (in Sect. 4 we show that we can achieve the same with much fewer plaintexts). Thus we obtain at most ω^{16} AES key candidates based only on constraints from round 1 one of which is the correct key. When $\omega = 8$ we get $\omega^{16} = 2^{48}$.

Now we can use the detected leakage from round 2 to identify the correct key and discard the false ones. For a fixed plaintext and a given key candidate, we can deterministically compute the 2nd round key and the state at the end of round 1. We can then deduce the 16 SBox cells that are accessed in round 2 and compare them to the access pattern measured by the detector. The probability of matching the detected pattern is $\omega^{16}/2^{128}$. Therefore, for the ω^{16} candidates from round 1, we can expect $\approx \omega^{32}/2^{128}$ candidates to fit the leakage from both rounds. For $\omega = 8$ we get $\approx 2^{96}/2^{128} \ll 1$, so it is very likely that we will find just the single correct key.

Note that the above process is a naive method used only to illustrate that the leakage from the first two AES rounds is sufficient to uniquely identify the correct key. However, we can do much better: We devised a specialized solver that has a time complexity of 2^{20} and space complexity of 2^{23} bits, when $\omega = 8$.

3.2   The Solver

Let a *partial key* be an array of 16 cells, each of which may contain either a value 0...255 or 'undefined'. The main algorithm maintains a set of partial key candidates, and works in stages. Each stage corresponds to a particular state byte, or a set of state bytes, in round 2: In the stage for state byte j the algorithm first grows the set of candidates, by extending each candidate partial key so all the key bytes that state byte j depends on are well defined. Then the algorithm rejects all the (extended) candidates that are inconsistent with 2nd round leaks. A stage can correspond to several state bytes if the extended candidate keys are well defined for all the depended-upon key bytes of the stage. The pseudo-code for a single stage has the following structure:

```
//stage for state byte j
input: set prevCandidates
Let enumBytes(j) be the set of additional key bytes that state byte j depends on and
are still 'undefined' in all partial keys in prevCandidates.
1: for all C in prevCandidates do
2:    for all possible values V for key bytes in enumBytes(j) do
3:        if Consistent (j, C||V) then
4:            nextCandidates ← nextCandidates ∪ {C||V}
5:        end if
6:    end for
7: end for
8: prevCandidates ← nextCandidates
9: nextCandidates = ∅
```

We keep the results of the 2nd round row activations in a data structure denoted by R2A: R2A$\{p^t\}$ is a vector of L bits such that $(\text{R2A}\{p^t\})_j = 1$ if plaintext p^t caused a detectable SBox access in round 2 on state byte j.

For a given partial key X and state byte $1 \le j \le 16$ line 3 calls a function to test whether X is consistent with the 2nd round leaks for state byte j:

```
1: Consistent (j, X)
2: for all plaintexts p^t do
3:    v_jt ← RowLookupOf (j, X, p^t)
4:    if ((v_jt == d and (R2A{p^t})_j==0) or (v_jt != d and (R2A{p^t})_j==1)) then
5:        return FALSE //partial key X is inconsistent
6:    end if
7: end for
8: return TRUE //partial key X is consistent
```

The function RowLookupOf (j, X, p^t) at line 3 returns the SBox row that is looked up for state byte j with plaintext p^t and partial key X. We ensure that all the key bytes that state byte j depends on are well defined in X by a careful ordering of the enumeration (see below), that also ensures the algorithm's ability to disqualify partial keys early. The time complexity of Consistent (j, X) is clearly $O(N_p)$, where N_p is the number of plaintexts.

Table 1. The algorithm going over bytes of the second round state matrix column by column. For every stage of the solver the number of candidates increases due to the newly enumerated key bytes—but the number of remaining candidates after the stage is reduced due to the second round constraints. This analysis assumes one second round activation for each of the state matrix byte j, and $\omega = 8, L = 32$, thus each stage cuts down the number of candidates by a factor of $\sim 2^5$.

Stage	Column	State byte	Bytes enumerated	Candidates	Running complexity	Space (bits)
1	1	1	$1, 6, 11, 14, 16$	2^{15}	2^{18}	$5 \cdot 2^3 \cdot 2^{15}$
2	1	3	3	$2^{10} \cdot 2^3$	2^{16}	$6 \cdot 2^3 \cdot 2^{13}$
3	1	2	$2, 15$	$2^8 \cdot 2^6$	2^{17}	$8 \cdot 2^3 \cdot 2^{14}$
4	1	4	$4, 13$	$2^9 \cdot 2^6$	2^{18}	$10 \cdot 2^3 \cdot 2^{15}$
5	2	$5, 6$	$5, 10$	$2^{10} \cdot 2^6$	2^{19}	$12 \cdot 2^3 \cdot 2^{16}$
6	2	7	7	$2^6 \cdot 2^3$	2^{12}	$13 \cdot 2^3 \cdot 2^9$
7	2	8	8	$2^4 \cdot 2^3$	2^{10}	$14 \cdot 2^3 \cdot 2^7$
8	3	$9, 10, 11$	9	$2^2 \cdot 2^3$	2^8	$15 \cdot 2^3 \cdot 2^5$
9	3	12	12	$2^{-10} \cdot 2^3$	2^{-4}	$16 \cdot 2^3$
10	4	$13, 14, 15, 16$	-	2^{-27}	2^{-24}	$16 \cdot 2^3$

3.3 Selecting the Enumeration Order

According to the appendix, state byte 1 depends on key bytes $1, 6, 11, 16$ after the round 1 MixColumns step, and byte 1 of round key 2 depends on key bytes $1, 14$. Thus immediately before the SBox lookup of round 2, state byte 1 depends on 5 key bytes: $1, 6, 11, 14, 16$ (see Fig. 2a). So in the solver's stage 1 we enumerate over a set of ω^5 candidates. Roughly speaking when a 2nd round row activation is detected for state byte 1, the consistency check will reduce the set to about $\frac{\omega^5}{L} \approx 2^{10}$ candidates. In the same way we find that state byte 3 depends on key bytes $1, 3, 6, 11, 16$—4 of which we've already enumerated in stage 1 (see Fig. 2b). So we only need to extend each candidate partial key by a single byte. Thus we enumerate on byte 3 for the second stage. After this stage (assuming 2nd round activation for the corresponding state byte) the number of candidates becomes $\approx (\frac{\omega^5}{L}) \cdot \omega \cdot \frac{1}{L} = \frac{\omega^6}{L^2}$, which is 2^8 when $\omega = 8$.

Continuing in a similar manner, we find that state byte 2 depends on 6 key bytes: $1, 2, 6, 11, 15, 16$ so we need to extend the partial keys by 2 bytes (2 and 15), ending the stage with $\frac{\omega^8}{L^3} = 2^9$, and so forth column by column. Table 1 illustrates the whole process. The figure shows that stage 5 dominates the time complexity (of 2^{19}) and space complexity (of 2^{21}) for a total time complexity of $\approx 2^{20}$.

Note that the state bytes of the first column (state bytes 1–4) collectively depend on 10 key bytes. A simpler algorithm would have enumerated over all 10 bytes together. However, such an approach would have had a time complexity of $\omega^{10} = 2^{30}$ (for $\omega = 8$)—significantly worse than the time complexity of our stages 1–4 combined.

$$(a) \qquad\qquad\qquad (b)$$

$$\begin{bmatrix} k_1 & \cdot & \cdot & * \\ * & k_6 & \cdot & k_{14} \\ * & \cdot & k_{11} & * \\ * & \cdot & \cdot & k_{16} \end{bmatrix} \qquad \begin{bmatrix} k_1 & \cdot & \cdot & * \\ * & k_6 & \cdot & * \\ k_3 & \cdot & k_{11} & * \\ * & \cdot & \cdot & k_{16} \end{bmatrix}$$

Fig. 2. The key bytes affecting the round 2 SBox accesses: (a) for state byte 1, (b) for state byte 3. Note that the key bytes on the diagonal $(1, 6, 11, 16)$ determine the state bytes of the 1st column at the end of round 1, and the key bytes on the left and right columns determine the 2nd round key.

4   Choosing the Plaintexts

As stated in Sect. 3 when a row access is detected in round 1, the number of key candidates for that byte is reduced to ω. The SBox values are located over L sequential rows of the SRAM memory, so the probability to observe a row access for randomly chosen plaintext is $\approx 1/L$.

For the set of plaintexts $p^t = (a_t, \dots, a_t)$ we use, we want to have at least one detectable row access in round 1 for every key byte. This can of course be guaranteed by using all 256 plaintexts, as done by [15]. However we can achieve the same result with much fewer plaintexts. For a given offset (recall Fig. 1(a)), a plaintext byte a_t, and key byte k_j, the AES SubBytes step generates an SRAM row access to row l

$$l = \left\lfloor \frac{a_t \oplus k_j + \text{offset}}{\omega} \right\rfloor + 1 \tag{1}$$

We capitalize on this by using a "ω-jump" strategy for plaintext ordering. We choose the following plaintexts:

$$p^t = \{c + j \cdot \omega, \dots, c + j \cdot \omega\} \tag{2}$$

for $c = \{0, \dots, \omega - 1\}$, and $j = \{0, \dots, L - 1\}$ for offset $= 0$ or $j = \{0, \dots, L - 2\}$ for offset $\neq 0$. Essentially for every value of c this strategy holds

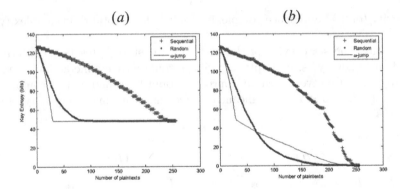

Fig. 3. The entropy of the key as function of the number of plaintexts, using only first round leakages for offset = 0 (a) and offset = 1 (b). The graphs show the sequential plaintext selection used in [15], a uniformly- random selection strategy and our "ω-jump" strategy. We can see that using only round-1 information, the entropy can't be reduced below 48 bit when offset = 0. We can see that using "ω-jump" the entropy decreases fast and using only 32 plaintexts we have a 48bit entropy, which is the "working point" of our solver, for all offsets.

the least-significant-bits fixed (e.g., the 3 LSBs for $\omega = 8$) and goes over all options for the MSBs.

By choosing some c and going over all options of j to multiply the row width ω we force a row access to all of the SRAM rows $\{1, 2, 3, \ldots, L\}$ for offset = 0 regardless of the key value k. If offset $\neq 0$, the "ω-jump" strategy causes a detectable row access for all the rows $\{2, 3, \ldots, L - 1\}$ plus one more row access— to the first or the last row depending on the chosen value of c. After going over all the values of j we increment c and repeat. By setting the detectable row d to be $2 \leq d \leq L - 1$ and using a set of L (or $L - 1$) plaintexts of Eq. (2) we are guaranteed to have one detectable row activation for every key byte during the first AES round. Figure 3 shows the drop in key entropy as a function of the number of plaintexts. Figure 3(b) shows that for offset = 1 the random strategy of plaintexts selection reduces the entropy to 0 quicker than the "ω-jump" strategy, but using the "ω-jump" strategy the entropy reaches the desired working point of our solver (48 bit entropy) using only L carefully chosen plaintexts.

Note that unlike the first round, the second round row activations can't be controlled by the choice of plaintexts since the access pattern in round 2 also depends on the key diffusion caused by round 1.

5   Decoding the Photonic Traces with Auto Threshold Calibration

For each of the plaintexts p^t we activate the IC (or, in our case, the simulator) M times. For each activation we count the number of detected photons per time step, while the detector is fixed at SRAM row d. We summarize the detection counts per time step, to obtain a "photonic trace" $T(p^t)$ for each plaintext, for the time duration of the first 2 AES rounds. Following [14,15] we assume an IC instruction cycle of 800 ns[1], a photonic trace spans 25.6 µs, represented by a vector of 1280 samples, one per 20 ns. For plaintext p^t we now need to decode the trace to extract two arrays of 16 bits: R1A and R2A recording the results of the 2 AES rounds' SBox activations. A bit value of 1 indicates that the plaintext caused a detectable SRAM access on the current SBox activation. A natural decoding rule is to use a threshold: if the number of detected activations during SBox access j in round 1 exceeds the threshold, we set $(R1A \{p^t\})_j = 1$, and 0 otherwise, and similarly for R2A.

A crucial task is calibrating the threshold so it can reliably distinguish between true detections and noise. Calibrating a threshold is often a heuristic trial-and-error process. However, since we choose the plaintexts in a specific way, we can calibrate the threshold automatically to its optimal value.

5.1   Calibration at High SNR

For illustration purposes we start by considering what happens when the SNR is high. Our method of choosing plaintexts guarantees a first round detectable row activation for every state byte j for at least one plaintext. Therefore we aggregate the N_p photonic traces (one per plaintext) by taking the *maximum* count per time step:

$$(maxT)_i = \max_{t=1...N_p} \left\{ (T(p^t))_i \right\} \qquad (3)$$

for the time duration of AES round 1.

This max-trace should exhibit 16 distinct peaks, at the time-steps corresponding to the 16 SBox activations of AES round 1. If we sort maxT in descending order, we expect to see a clear drop between the 16th peak value, and the 17th (which is the highest peak caused by the noise). We can use this fact and choose our threshold to be the midpoint between the two peaks:

$$Threshold = \frac{peak_{16} + peak_{17}}{2} \qquad (4)$$

where $peak_{16}$ and $peak_{17}$ are the 16th and 17th largest samples of $maxT$.

Even though the threshold is calibrated on $maxT$ for the first AES round, it is valid for every individual trace $T(p^t)$, and for both AES rounds. Thus we can use this threshold for all N_p traces to set the bit arrays $R1A \{p^t\}$ and $R2A \{p^t\}$.

However, we do not use this basic calibration. Instead, in the next section we show a more delicate calibration with two thresholds, that converge to the basic threshold when SNR is high.

---

[1] Note that this clock frequency is a slow 1.25 MHz. The AT-Mega328p can operate at faster clock frequencies, up to 20 MHz- we simulated the 1.25 MHz clock to allow a comparison of the simulated results with the findings of [14,15].

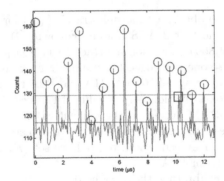

Fig. 4. A trace and the low and high thresholds for M = 1,000,000 (low SNR). In circles, peaks at expected time slots. In a box, a peak at an unexpected time slot. Thus, Thr_1 is set just below the lowest circled value, and Thr_2 is set just above the boxed value.

5.2   Calibration at Low SNR

When the number of measurements M for each plaintext is low, the SNR drops and the threshold calibration method of Sect. 5.1 starts to introduce decoding errors. We can define 2 error types:

1. False negative: a missed row activation (threshold was set too high).
2. False positive: an incorrect row activation (threshold was set too low).

We separate the discussion of the errors into two cases, for the first and second rounds of the AES process.

Recall that our solver (Sect. 3.2) uses the first AES round activations to reduce the number of candidates from 256 to ω for every key byte. When a false positive occurs during the first AES round we will have more than ω options for the key byte, since we will have ω options for each activation. This could make the solver running time slower and cause the set of final key candidates to be larger. However, when a false negative occurs during the first AES round, we are left with 256 options for this key byte. Since the key bytes options are used to enumerate over all key options, too many options can make the solver running time unaffordable. Thus in AES round 1 we prefer to set the threshold low, and suffer occasional false positives.

Second AES round activations set constraints that the solver uses to disqualify key candidates obtained from first round leakages. A false negative during the second round would cause fewer constraints and weaker disqualifications—so the solver may end with more keys. However, a false positive would disqualify true key values. Therefore in AES round 2 we prefer to set the threshold too high, and suffer occasional false negatives.

Our solution is to use two thresholds: one for each AES round. The first threshold (Thr_1) is set low in order to avoid false negative errors of first round activations. The second threshold (Thr_2) is set higher in order to avoid second round false positives. To calibrate the thresholds we again use the max-trace

$maxT$. We utilize the fact that we know the time-steps in which the 16 S-Box accesses occur. We use the following process to calibrate the two thresholds.

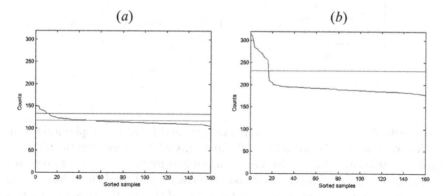

Fig. 5. The sorted maxT trace and the auto-calibrated thresholds (lines) for (a) M = 1,500,000 and (b) M = 3,750,000 measurements. We can notice on (b) a gap between the 16th and the 17th samples and the two thresholds converge.

1. Generate the max-trace $maxT$ as in Sect. 5.1.
2. Thr_1 is set to the maximal value for which $(maxT)_i \geq \text{Thr}_1$ for all 16 time-steps i at which there is a first round activation.
3. Thr_2 is the minimal value for which $(maxT)_i < \text{Thr}_2$ for all time-steps i at which there is *no* first-round activation (see Fig. 4).

If the SNR is high then peaks at the 16 true activations will be all higher than the noise—so we will get $\text{Thr}_1 \geq \text{Thr}_2$. In such a case we fall back to the method of Sect. 5.1 and set both thresholds to be $(\text{Thr}_1 + \text{Thr}_2)/2$ (see Fig. 5).

We take key candidates based on first round activations using Thr_1, and we collect the constraints from the second round activations using Thr_2.

6 Practical Results

We implemented the photonic emissions simulator in Matlab. The solver was implemented in python. The experiments were run on a relatively old Intel Core Duo T2450 2 GHz, 2 GB RAM PC running Windows Vista. We simulated the ATmega328P IC with SRAM row width of $\omega = 8$ and generated the plaintexts according to the "ω-jump" strategy of Sect. 4.

In order to evaluate the performance of our attack we performed an extensive set of experiments. All the experiments were done with $\omega = 8$, and with either $L = 32$ (for offset = 0) or $L = 33$ (for all other offsets). We used the "ω-jump" strategy to generate L plaintexts for each offset.

Table 2. A comparison between the SPEA and our E-SPEA methods.

Method	Final key candidates		Plaintexts	Measurements per plaintext	Total measurements	Time (hours)
SPEA	$\begin{cases} 1 \\ 2^{48} \\ 2^{32} \\ 2^{16} \end{cases}$	for odd offset for offset $= 0$ for offset $= 4$ for offset $= 2, 6$	256	5 M	1280 M	6.4
E-SPEA	$\begin{cases} 1 \\ \sim 8 \\ 2^{48} \end{cases}$	for 75 % for 24 % for 1 %	32	1.5 M	48 M	0.5

For each plaintext we used 100 random keys, and for each key-plaintext combination we generated between $M = 1,000,000$–$5,000,000$ traces from the photonic emission simulator, with the detector at a random row $2 \leq d \leq L - 1$. We used the threshold setting of Sect. 5 to decode the traces, and used the solver to find the key. For each run we set a timer on the solver: if the run time exceeded $5000\,s$ we stopped it and recorded a failure. Figure 6 shows the attack's behavior for various values of M. We can see that as long as $M \geq 1,500,000$ the attack works well, with the median key entropy at the end of the attack dropping below 3 bits, and a single (correct) key was found in 75 % of the runs. When $M \geq 1,500,000$ the attack takes under 10 min, on our slow PC. The results for other offsets were similar (graphs omitted).

Table 2 shows a comparison of our Enhanced SPEA with the original SPEA, and Fig. 7 shows the running time of solver. The Table shows that due to the reduced number of required plaintexts, and reduced number of required measurements M, our total attack time drops by an order of magnitude, from 6.4 h down to 30 min- while succeeding in finding a single (correct) key in 75 % of the

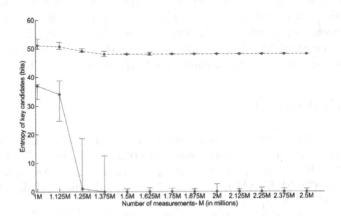

Fig. 6. The entropy of the round-1 key candidates (dashed line) and the final key candidates (solid line) as a function of the number of measurements M, for offset $= 0$ and using different random keys and a different detector row for every test. The upper and lower bounds indicate the 5–95 percentiles and the dots mark the median values.

cases- regardless of the offset. The E-SPEA method however had difficulty with 1 % of the cases, not getting below 2^{48} key candidates: in those cases the number of second round activations was very low and the solver reached a timeout of 5000 s without being able to reduce the number of key candidates.

7   Conclusions, Future Work and Countermeasures

In this paper we demonstrated that using cryptographic post-processing, careful plaintext selection, and better signal processing, we are able to significantly improve upon the SPEA of [15]. We are able to uniquely extract the correct key regardless of the offset at which the SBox is placed in SRAM. We achieve this while reducing the required number of photonic measurements by an order of magnitude, which directly implies a similar drop in the attack's time complexity. Our cryptographic solver is extremely efficient, with a time complexity of 2^{20}, and extracts the key within minutes on a rather old PC.

Following [15] we evaluated our attack assuming an SRAM row width of $\omega = 8$, as in the ATMega328P. However, we note that a row width of $\omega = 16$ (as in the ATXMega128A1) would pose a harder challenge: we expect to find $\approx \omega^{32}/2^{128} = 1$ key candidates that fit the leakage from the first two AES rounds, as opposed to the $\approx 2^{-32}$ expected when $\omega = 8$. I.e., in the intermediate stages we will have many more key candidates, the run time will be longer, and the attack will terminate with more possible keys, than when $\omega = 8$. Conversely, if $\omega = 32$ then our attack should become equally efficient as when $\omega = 8$: we can set the detector on the *column*-access transistor. We leave evaluating alternative SRAM configurations for future work.

Note also that our photonic emissions simulator allows us to test hypothetical lab setups, since we can experiment with the lens area and height above the IC,

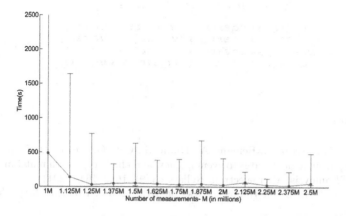

Fig. 7. Solver running time for different M values, for offset = 0 and using different random keys and a different detector row for every test. The upper and lower bounds indicate the 5–95 percentiles and the dots mark the median values.

the supply voltage, the temperature, and the detector sensitivity. It would be interesting to use the simulator's results to guide the design of better future detectors.

The attack is susceptible to countermeasures such as delays and dummy operations which can obfuscate the time a photonic emission may occur. Masking also can make the attack more difficult. Memory protection countermeasures such as memory encryption or scrambling have no effect on the emission pattern, but they can make the preliminary stage of finding the SBox values inside the SRAM memory more difficult.

Appendix

The AES Process Until the Second SubBytes Operation

References

1. Bascoul, G., Perdu, P., Benigni, A., Dudit, S., Celi, G., Lewis, D.: Time resolved imaging: from logical states to events, a new and efficient pattern matching method for VLSI analysis. Microelectron. Reliab. **51**(9), 1640–1645 (2011)
2. Bernstein, D.J.: Cache-timing attacks on AES (2004). Preprint, http://cr.yp.to/papers
3. Bertoni, Y.M., Grassi, L., Melzani, F.: Simulations of optical emissions for attacking AES and masked AES. In: Chakraborty, R.S., Schwabe, P., Solworth, J. (eds.) Security, Privacy, and Applied Cryptography Engineering (SPACE). LNCS, vol. 9354, pp. 172–189. Springer, Verlag (2015)

4. Carmon, E., Seifert, J.-P., Wool, A.: Simple photonic emission attack with reduced data complexity. Cryptology ePrint Archive, Report 2015/1206 (2015). http://eprint.iacr.org/2015/1206
5. Chynoweth, A., McKay, K.: Photon emission from avalanche breakdown in silicon. Phys. Rev. **102**(2), 369 (1956)
6. Di-Battista, J., Courrege, J.C., Rouzeyre, B., Torres, L., Perdu, P.: When failure analysis meets side-channel attacks. In: Mangard, S., Standaert, F.X. (eds.) CHES 2010. LNCS, vol. 6225, pp. 188–202. Springer, Heidelberg (2010)
7. Egger, P., Grützner, M., Burmer, C., Dudkiewicz, F.: Application of time resolved emission techniques within the failure analysis flow. Microelectron. Reliab. **47**(9), 1545–1549 (2007)
8. Ferrigno, J., Hlavác, M.: When AES blinks: introducing optical side channel. Inf. Secur. **2**(3), 94–98 (2008)
9. Krämer, J., Kasper, M., Seifert, J.-P.: The role of photons in cryptanalysis. In: 19th Asia and South Pacific, Design Automation Conference (ASP-DAC), pp. 780–787. IEEE (2014)
10. Krämer, J., Nedospasov, D., Schlösser, A., Seifert, J.P.: Differential photonic emission analysis. In: Prouff, E. (ed.) COSADE 2013. LNCS, vol. 7864, pp. 1–16. Springer, Heidelberg (2013)
11. Nedospasov, D., Seifert, J.-P., Schlosser, A., Orlic, S.: Functional integrated circuit analysis. In: IEEE International Symposium on Hardware-Oriented Security and Trust (HOST), pp. 102–107. IEEE (2012)
12. Newman, R.: Visible light from a silicon pn junction. Phys. Rev. **100**(2), 700–703 (1955)
13. Osvik, D.A., Shamir, A., Tromer, E.: Cache attacks and countermeasures: the case of AES. In: Pointcheval, D. (ed.) CT-RSA 2006. LNCS, vol. 3860, pp. 1–20. Springer, Heidelberg (2006)
14. Schlösser, A.: Hot electron Luminescence in silicon structures as photonic side channel (in German). Ph.D. thesis, Faculty of Mathematics and Natural sciences, Berlin Institute of Technology (2014)
15. Schlösser, A., Nedospasov, D., Krämer, J., Orlic, S., Seifert, J.-P.: Photonic emission analysis of AES. In: Workshop on Cryptographic Hardware and Embedded Systems (CHES) (2012)
16. Schlösser, A., Nedospasov, D., Krämer, J., Orlic, S., Seifert, J.-P.: Simple photonic emission analysis of AES. J. Cryptographic Eng. **3**(1), 3–15 (2013)
17. Selmi, L., Mastrapasqua, M., Boulin, D.M., Bude, J.D., Pavesi, M., Sangiorgi, E., Pinto, M.R.: Verification of electron distributions in silicon by means of hot carrier luminescence measurements. IEEE Trans. Electron Devices **45**(4), 802–808 (1998)
18. Song, P., Stellari, F., Huott, B., Wagner, O., Srinivasan, U., Chan, Y., Rizzolo, R., Nam, H., Eckhardt, J., McNamara, T., et al.: An advanced optical diagnostic technique of IBM z990 eserver microprocessor. In: Proceedings IEEE International Test Conference (ITC), p. 9. IEEE (2005)
19. Weste, N., Harris, D., Design, C.: A Circuits And Systems Perspective, 4/E. Pearson Education, (2010)

Side-Channel Analysis (Case Studies)

Power Analysis Attacks Against IEEE 802.15.4 Nodes

Colin O'Flynn[✉] and Zhizhang Chen

Dalhousie University, Halifax, Canada
{coflynn,zchen}@dal.ca

Abstract. IEEE 802.15.4 is a wireless standard used by a variety of higher-level protocols, including many used in the Internet of Things (IoT). A number of system on a chip (SoC) devices that combine a radio transceiver with a microcontroller are available for use in IEEE 802.15.4 networks. IEEE 802.15.4 supports the use of AES-CCM* for encryption and authentication of messages, and a SoC normally includes an AES accelerator for this purpose. This work measures the leakage characteristics of the AES accelerator on the Atmel ATMega128RFA1, and then demonstrates how this allows recovery of the encryption key from nodes running an IEEE 802.15.4 stack. While this work demonstrates the attack on a specific SoC, the results are also applicable to similar wireless nodes and to protocols built on top of IEEE 802.15.4.

Keywords: AES · Side-channel power analysis · DPA · IEEE 802.15.4

1 Introduction

IEEE 802.15.4 is a low-power wireless standard which targets Internet of Things (IoT) or wireless sensor network (WSN) applications. Many protocols use IEEE 802.15.4 as a lower layer, including ZigBee (which encompasses many different protocols such as ZigBee IP and ZigBee Pro), WirelessHART, MiWi, ISA100.11a, 6LoWPAN, Nest Weave, JenNet, IEEE 802.15.5, Thread, Atmel Lightweight Mesh, and DigiMesh. As part of the IEEE 802.15.4 standard a security suite based on AES is included, which allows encrypting and adding an authentication code on the wireless messages.

Protocols using IEEE 802.15.4 as a lower layer often include security at layers above IEEE 802.15.4, but many of them also use the same AES primitive as the lower layer (with a different key and possibly encryption mode). An attack against the AES peripheral in an embedded device may be useful in attacking both the lower and higher layers depending on network specifics. Even if acquiring the 802.15.4-layer key is not directly useful, because for example each link uses a different key, an attacker may practically benefit from the ability of sending arbitrary messages which will be accepted as valid and passed to the higher-layer protocol decoder logic. With this ability an attacker can exploit security flaws in higher-layer protocol decoding logic, since the lower-layer messages will be successfully decrypted and presented to higher layers.

F.-X. Standaert and E. Oswald (Eds.): COSADE 2016, LNCS 9689, pp. 55–70, 2016.
DOI: 10.1007/978-3-319-43283-0_4

This paper presents an attack against a wireless node that uses the IEEE 802.15.4 protocol. We present the following important results from developing this attack: (1) an attack against the hardware AES engine in the Atmel ATMega128RFA1, (2) an attack on AES-128 in CCM* mode as used in IEEE 802.15.4 [1], (3) a method of causing the AES engine in the target device to perform the desired encryption, and (4) a shunt-based measurement method for devices with internal voltage regulators. This attack is validated with a hardware environment (shown in Fig. 1).

The attack demonstrated here uses side-channel power analysis [2], specifically a correlation-based attack [3]. We obtained the power measurements in this work by physically capturing a node and inserting a shunt resistor. In general, side-channel attacks can be performed with a noncontact electromagnetic (EM) probe instead, which does not require modification to the device [4]. The EM measurement typically achieves similar results to the resistive shunt [5,6].

It has previously been demonstrated that wireless nodes are vulnerable to side-channel power analysis when running AES-ECB in software [7]. This type of attack does not destroy the node under attack, and the node will continue to function during the attack. This makes detection more difficult: although a node is captured, it still appears on the network. Our work extends this by attacking the actual AES-CCM* mode used in IEEE 802.15.4, attacking the hardware AES accelerators typically used in wireless stack implementations, and demonstrating how to force many encryption operations to occur for rapid collection of traces.

We begin by describing the attack on the ATMega128RFA1 AES hardware peripheral in Sect. 2. Next, we look at specifics of the use of AES encryption on the IEEE 802.15.4 wireless protocol in Sect. 3. This outlines the challenges of applying the side-channel attack to the AES-CCM* mode of operation, which is solved for the case of IEEE 802.15.4 in Sect. 4. Our application of this to a real IEEE 802.15.4 node is discussed in Sect. 5, and our conclusions follow.

An extended version of this paper is available which contains additional details and discussion of this attack[1].

2   ATMega128RFA1 Attack

The Atmel ATMega128RFA1 is a low-power 8-bit microcontroller with an integrated IEEE 802.15.4 radio, designed as a single-chip solution for Internet of Things (IoT) or wireless sensor network (WSN) applications [8]. As part of the IEEE 802.15.4 radio module a hardware AES-128 block is available, designed to work with the AES security specification of IEEE 802.15.4. Other vendors such as Freescale (MC13233), Silicon Laboratories (EM35x), STMicroelectronics (STM32W108), and Texas Instruments (CC2530) provide similar chips integrating an IEEE 802.15.4 radio and microcontroller in a single device.

To perform a side-channel power analysis attack, we evaluate a method of physically measuring power on the ATMega128RFA1 in Sect. 2.1. We then determine an appropriate power model in Sect. 2.2, and we present the results of the

---

[1] The extended version is published at https://eprint.iacr.org/2015/529.

Fig. 1. The ChipWhisperer capture hardware is used in this attack, along with details of the measurement point.

CPA attack [3] in Sect. 2.3. We present additional considerations for attacking intermediate rounds (i.e., beyond the first round) of the AES algorithm in Sect. 2.4; these intermediate-round attacks are required for the AES-CCM* attack.

2.1 Power Measurement

Power measurements can be performed by inserting a resistive shunt into the power supply of the target device, and measuring the voltage drop across the shunt. Because devices often have multiple power supplies (such as VCC_{core}, VCC_{IO}, VCC_{RF}), the shunt must be inserted into the power supply powering the cryptographic core. As with many similar IEEE 802.15.4 chips, the core voltage of the ATMega128RFA1 is lower (1.8 V) than the IO voltage (typically 2.8–3.3 V) [8].

To avoid requiring an external voltage regulator for the lower core voltage, most of these devices also contain an integrated 1.8 V voltage regulator. Some devices require an external connection from the regulator output pin to the VCC_{core} pin. With this type of device we could perform the power measurements by either (a) inserting a shunt resistor between the output and input, or (b) using an external low-noise power supply with a shunt resistor (as in [7]). The ATMega128RFA1 is not such a device – it internally connects the regulator to the VCC_{core} pin, but does require a decoupling capacitor placed on the VCC_{core} pin (which also serves as the output capacitor for the voltage regulator).

By inserting a shunt resistor into the path of the decoupling capacitor, we can measure high-frequency current flowing into the VCC_{core} pin. Note that this measurement will be fairly noisy, as we will also have noise from current flowing out of the voltage regulator. The right side of Fig. 1 shows the implementation of this arrangement. Externally powering this pin with a voltage slightly higher than 1.8 V may disable the internal regulator, giving a lower-noise signal from the shunt resistor. This is dependent on regulator design.

2.2 Related Hardware Attack

We based our work on Kizhvatov's attack on the XMEGA device [9]. Kizhvatov determined that for a CPA attack on the XMEGA, the Hamming distance between successive S-box input values leaked. These input values are the XOR of the plaintext with the secret key that occurs during the first AddRoundKey.

Our notation considers p_i and k_i to be a byte of the plaintext and encryption key respectively, where $0 \le i \le 15$. To determine an unknown byte k_i, we first assume we know a priori the value of p_i, p_{i-1}, and k_{i-1}.

This allows us to perform a standard CPA attack, where the sensitive value is given by the Hamming weight of (1). That is to say the leakage for unknown encryption key byte i is: $l_i = HW(b_i)$. Provided k_0 is known, this attack can proceed as a standard CPA attack, with only 2^8 guesses required to determine each byte.

$$b_i = (p_{i-1} \oplus k_{i-1}) \oplus (p_i \oplus k_i), \quad 1 \le i \le 15 \tag{1}$$

For the specific case of k_0, the Hamming distance from the fixed value 0x00 is used as a leakage model[2], as in (2). This allows the entire encryption key to be attacked with a total of 16×2^8 guesses.

$$l_0 = HW(b_0) = HW(p_0 \oplus k_0) \tag{2}$$

2.3 Application to ATMega128RFA1

Our experimental platform was a Dresden Elektronik radio board, model number RCB128RFA1 V6.3.1. To sample the power measurements, we used an open-source platform called the ChipWhisperer Capture Rev2 [10]. This capture hardware synchronizes its sampling clock to the device clock, and we configured it to sample at 64 MS/s (which is 4 times the ATMega128RFA1 clock frequency of 16 MHz). The differential probe is connected across a shunt in the VCC_{core} power pin as described previously. A filter with a passband of 3–14 MHz was inserted between the output of the differential probe and the low-noise amplifier input of the ChipWhisperer.

We implemented a test program in the ATMega128RFA1 that encrypts data received over the serial port. This encryption can be done via either a software AES-128 implementation or the hardware AES-128 peripheral in the ATMega128RFA1. When using the hardware peripheral, the encryption takes 25 μs to complete, or about 400 clock cycles.

We used a CPA attack, ranking the most likely byte as the one with the highest correlation values [3]. We use a plot of the partial guessing entropy (PGE) compared to number of traces in order to measure attack success [11]. The PGE indicates where the correct value of the encryption subkey byte falls within a list ordered from most to least likely based on CPA attack results.

---

[2] This is not published in [9], but was described in private communication from the author.

Thus when the PGE falls to zero the specific subkey byte is perfectly known, and a PGE of 128 would be expected for a completely unsuccessful attack that is equivalent to a random guess.

To evaluate our measurement toolchain, we performed this attack against a software AES implementation on the ATMega128RFA1, which recovered the complete key in under 60 traces.

We then recorded a total of 50 000 power traces, where the ATMega128RFA1 was performing AES-128 ECB encryptions using random input data during the time each power trace was recorded. For each trace, 600 data points were recorded at a sampling rate[3] of 64 MS/s. Each trace therefore covered about the first third of the AES encryption.

Our initial CPA attack was repeated five times over groups of 10 000 traces. The resulting average partial guessing entropy for each byte is shown in Fig. 2. The first byte (which uses the leakage assumption of (2)) has the worst performance, as the guessing entropy does not reach zero with 10 000 traces.

Guessing of k_{i-1}. This attack used the leakage (2) of the first byte $i = 0$ to bootstrap the key recovery. Once we know this byte, we can use (1) to recover successive bytes.

Practically, we may have a situation where $i - 1$ is not recoverable. Previous work assumed either some additional correlation peak allowing us to determine $i - 1$, or the use of a brute-force search across all possibilities of the byte $i - 1$ [9]. We can improve on this with a more efficient search algorithm, described next.

The leakage function (1) could be rewritten to show more clearly that the leaked value depends not on the byte values, but on the XOR between the two successive bytes, as in (3).

$$b_i = (k_{i-1} \oplus k_i) \oplus (p_{i-1} \oplus p_i), \quad 1 \leq i \leq 15 \qquad (3)$$

The side-channel attack can be performed with the unknown byte k_{i-1} set to 0x00, and the remaining bytes are recovered by the CPA attack described previously. These recovered bytes are not the correct value, but instead provide the value that has to be XOR'd with the previous byte to generate the correct byte.

The 256 candidate keys can then be generated with almost no computational work, by iterating through each possibility for the unknown byte k_{i-1}, and using the XOR values recovered from the CPA attack to generate the remaining byte values $k_i, k_{i+1}, \cdots, k_I$.

This assumes we are able to directly test those candidate keys to determine which is the correct value. As is described in the next section, we can instead use a CPA attack on the next-round key to determine the correct value of k_{i-1}.

---

[3] Note that this 64 MS/s sample rate is successful because the capture hardware samples synchronously with the device clock. If using a regular oscilloscope with an asynchronous timebase we expect a much higher sample rate to be required, similar to that reported in the XMEGA attack.

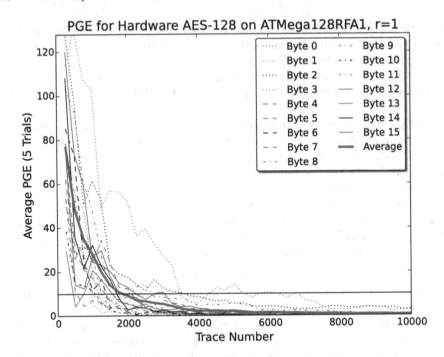

Fig. 2. The CPA attack on the hardware AES peripheral reduces the guessing entropy to reasonable levels in under 5000 traces, and is makes key recovery trivial in 10 000 traces. (Color figure online)

2.4 Intermediate-Round Attacks

Whereas our work so far has been concerned with determining the first-round encryption key, we will see in Sect. 4 that information on the round keys used during intermediate rounds is also required.

We determined that for intermediate rounds the leakage assumption of (1) and (2) still holds, where the unknown byte k_i is a byte of the round key, and the known plain-text byte p_i is the output of the previous round. We can extend our notation such that the leakage from round r becomes $l_i^r = HW(b_i^r)$, where each byte of the round key is k_i^r, and the input data to that round is p_i^r.

Examples of the PGE when attacking the start of the third round ($r = 3$) are given in Fig. 3. The entropy change for all rounds tested ($r = 1, 2, 3, 4$) was similar.

For details of the execution time of the hardware AES implementation, refer to Table 1. This table shows the samples used for each byte in determining the most likely encryption key for the first four rounds. For byte 0 (the first byte), (2) is the sensitive operation. For later bytes (1) is the sensitive operation.

Note the sample rate is four times the device clock, and in Table 1 the sample delta from start to end of the sensitive operations within each round is about 64 samples, or 16 device clock cycles. This suggests that a sensitive operation

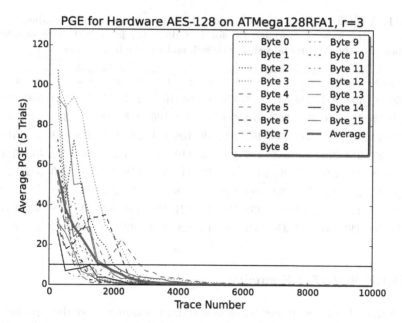

Fig. 3. Attacking intermediate rounds in the AES peripheral is also successful using the same leakage assumptions as the first-round attack. (Color figure online)

is occurring on each clock cycle. Each round takes approximately 32–34 cycles based on the repeating nature of the leakages in intermediate rounds.

Determining k_{i-1} Using Intermediate Rounds. As described in Sect. 2.3, we can perform the CPA attack on byte k_i where k_{i-1} is unknown by determining not the value of the byte, but the XOR of each successive byte with the previous key. This means performing the attack first where k_{i-1} is assumed to be 0x00.

By then enumerating all 256 possibilities for k_{i-1}, we can quickly generate 256 candidate keys to test. But if we are unable to test those keys, we need another way of validating the most likely value of k_{i-1}.

If we knew the initial (first-round) key, we could determine the input to the second round, and thus perform a CPA attack on the second-round key. Instead we have 256 candidates for the first round ($r = 1$), and want to determine which of those keys is correct before proceeding.

To determine which of the keys is correct, we can perform a CPA attack on the first byte of the second round, k_0^2, repeating the CPA attack 256 times, once for each candidate first-round key.

The correlation output of the CPA attack will be low for all guesses of k_0^2 where $\mathbf{k^1}$ is wrong, and only for the correct guess of k_0^2 and $\mathbf{k^1}$ will there be a peak. This technique will be used in Sect. 4.1, where we cannot test candidate keys as we are not recovering the complete key.

Table 1. A small range of points is selected from each trace, corresponding to the location of the device performing (2) for $i = 0$, or (1) for $i \geq 1$. The variable r corresponds to the AES round being attacked, and i is the byte number.

i	$r = 1$	$r = 2$	$r = 3$	$r = 4$	i	$r = 1$	$r = 2$	$r = 3$	$r = 4$
0	66–70	198–204	336–342	474–478	8	98–102	233–237	370–374	506–508
1	70–75	205–210	340–345	478–481	9	101–106	237–241	373–377	510–513
2	73–78	208–215	345–348	482–489	10	106–111	240–247	378–383	514–519
3	79–83	213–216	350–355	486–490	11	110–114	245–250	382–385	518–521
4	81–88	218–221	355–368	490–494	12	114–119	248–254	385–390	522–524
5	85–90	220–225	358–361	494–498	13	118–123	253–258	390–394	525–529
6	89–95	225–233	362–365	498–501	14	121–126	258–265	394–398	530–534
7	93–98	230–235	366–370	502–505	15	126–129	262–268	398–402	534–538

3   IEEE 802.15.4 Security

IEEE 802.15.4 is a low-power wireless standard, sending short data packets of up to 127 bytes at bit-rate of 250 kbit/s. The IEEE 802.15.4 standard uses AES-128 as the basic building block for both encryption and authentication of messages. The standard defines a mode of operation called CCM*, which modifies the regular CCM mode by allowing the use of encryption without authentication [1,12].

The underlying encryption uses AES-CTR mode, with an input format as shown in Fig. 4. The first 14 bytes are the nonce, and the last two bytes are the AES-CTR mode counter. Each received frame must use a new nonce, as the counter only counts the number of 16-byte blocks within the frame.

To ensure nonce freshness, a field called FrameCounter is included with each transmitted message and used as part of the nonce. The receiver verifies that the value of FrameCounter is larger than any previously used value, avoiding the reuse of a nonce.

On receiving a packet, the IEEE 802.15.4 layer first returns an acknowledgment to the sender. If the packet has security enabled (it is encrypted or has an authentication code appended) the node performs the following steps: (1) validates headers, (2) check the new received frame counter is numerically greater

0	1	2	3	4	5	6	7
Flags	Source Long Address						

8	9	10	11	12	13	14	15
Addr (cont'd)	FrameCounter				SecLevel	AES Counter	

Fig. 4. The following data is used as the input to AES-128 when a frame is decrypted by an IEEE 802.15.4 stack. The FrameCounter can be controlled by the attacker.

than the last valid frame count, (3) looks up the secret key based on addressing, (4) decrypts the payload and authentication code (if present), (5) validates the authentication code (if present), and (6) stores the frame counter.

For our side-channel attack we only care that step 4 is performed; this means our packet must successfully pass through steps 1–3. This requires that the packet is properly addressed and has an acceptable security configuration, i.e. using a valid key identifier and address. An example of such a packet is available in the extended version of this paper.

4   Application to AES-CCM* Mode

For a standard CPA attack, we require the ability to cause a number of encryption operations to occur with known plaintext or ciphertext material. In addition, the data being encrypted must vary between operations, as otherwise each trace will generate the same hypothetical intermediate values during the search operation of the CPA attack.

From Sect. 3 and Fig. 4, we know that a number of the bytes are fixed during the AES encryption operation. Practically *all* the bytes except for the FrameCounter are considered fixed in this attack. The Flags and SecLevel bytes will have constant (and known) values. Initially it would appear that the Source Long Address and AES Counter fields may vary, but as we discuss next, this is not the case.

The Source Long Address field comes from internal tables in the 802.15.4 stack, and is not simply copied from the incoming packet fields. The AES Counter field changes during operation, as it increases for each 16-byte block encrypted in AES-CCM* mode. But as the IEEE 802.15.4 packet is limited to a total of 127 bytes, the AES Counter field could never exceed 0x0007. Thus, between these 10 bytes, at most 3 bits vary during operation.

We instead rely on the ability of the attacker to control the FrameCounter field to mount a successful attack on an IEEE 802.15.4 wireless node. For our work we will assume an attack on the first encryption operation when a packet is received, meaning the AESCounter field is also fixed. The sent value of FrameCounter must simply be higher than a previously accepted value, which can either be determined by passive listening, or the most significant bit(s) can simply be set high to guarantee values which are likely to be accepted.

4.1   Previous AES-CTR Attacks

The AES-CCM* mode used by IEEE 802.15.4 is a combination of CBC-MAC and CTR modes of operation. Our attack is on the AES-CTR portion of the algorithm, with some modifications to reflect the use of a frame counter for the nonce material.

Previous work on AES-CTR mode has focused on the assumption that we can cause a number of encryptions to occur in sequence (i.e., with increasing counter number), but with unknown but constant nonce material [13]. Our work uses many of the constructs developed by Jaffe in [13], but with different assumptions of inputs on the AES block and a different leakage model. These differences necessitate the development of new techniques to recover partial keying information, as we cannot directly apply the previously published attack.

In our case, we have the ability to change 4 bytes of the input plaintext (bytes 9, 10, 11, and 12). The CPA attack only allows us to recover these four bytes of the key, as the keying material associated with bytes 9–12 can be recovered by a standard CPA attack using the leakage model identified in Sect. 2. The remaining bytes cannot be recovered, as the input data is constant, and hence our leakage target of the difference between S-Box inputs is also constant.

For the $MixColumns()$ operation, we can represent the four input bytes – one column of the state matrix – with $s_0, \cdots, s_3$, and the resulting output bytes with $S_0, \cdots, S_3$. The $MixColumns()$ operation uses multiplication over the Galois field GF(2^8), where we represent this multiplication operation with the symbol "$\circ$". The $MixColumns()$ operation then becomes:

$$S_0 = (2 \circ s_0) \oplus (3 \circ s_1) \oplus s_2 \qquad \oplus s_3 \tag{4}$$

$$S_1 = s_0 \qquad \oplus (2 \circ s_1) \oplus (3 \circ s_2) \oplus s_3 \tag{5}$$

$$S_2 = s_0 \qquad \oplus s_1 \qquad \oplus (2 \circ s_2) \oplus (3 \circ s_3) \tag{6}$$

$$S_3 = (3 \circ s_0) \oplus s_1 \qquad \oplus s_2 \qquad \oplus (2 \circ s_3) \tag{7}$$

Using the method from [13], we use our partial knowledge of the current round key to recover information about the next round key. Performing the attack with partial knowledge is possible as if some of the input bytes to $MixColumns()$ are fixed but unknown, we set those fixed bytes to 0, and use the linear property of $MixColumns()$ to introduce a correction constant. Assuming the true output of one $MixColumns()$ is S_0, we define the output that results by setting constant bytes to 0 as $S_0' = S_0 \oplus E_0$, where E_0 is an unknown correction constant.

Performing the CPA attack using the assumed output S_0', we would recover a version of this round key byte (we will refer to it as k_0') XOR'd with the unknown constant E_0, that is $k_0' = k_0 \oplus E_0$. The output of $AddRoundKey()$ will be equivalent to the case where we had the true key and true input, as:

$$AddRoundKey(k_0', S_0') = k_0' \oplus S_0' = (k_0 \oplus E_0) \oplus (S_0 \oplus E_0) = k_0 \oplus S_0 \tag{8}$$

This is sufficient information to perform the attack on the next round of the AES algorithm. Thus, if the entire modified version of a key can be recovered for a given encryption round, we can recover the entire *unmodified* key by attacking the next encryption round. This unmodified key can then be rolled backwards using the AES key schedule.

Description of Attack. We describe the attack by working through a symbolic example, using the following variables:

p_i^r : "text" input to $AddRoundKey()$ X : variable and known inputs
k_i^r : "key" input to $AddRoundKey()$ Y : variable and known intermediates
E_i^r : a constant, see Sect. 4.1   Z : variable and known intermediates
n_i^r : the modified round key, $k_i^r \oplus E_i^r$ c : constant values
s_i^r : the output of $SubBytes()$? : variable and unknown values
v_i^r : the output of $ShiftRows()$ N : known modified round-key values (n_i^r)
m_i^r : the output of $MixColumns()$ K : known key or round-key values (k_i^r)
X* : group of variables which has a small set of candidates for the correct value

Initially, we have the known input plaintext, where 12 of the bytes are constant, and the 4 variable bytes are under attacker control (FrameCounter). From this, we can perform a CPA attack to recover 4 bytes of the key. Note that in practice the byte k_9^1 cannot be recovered because k_8^1 is unknown. Instead we use the technique detailed in Sect. 2.4 to generate 256 candidate keys for $k_9^1, \cdots, k_{12}^1$, and test them at a later step. This means we can assume the following is the state of our initial-round key:

k^1 = [c c c c c c c c c K*K*K*K* c c c]

This can be used to calculate the output of the $SubBytes()$ and $ShiftRows()$ functions, where the majority of bytes are constant (but unknown):

s^1 = [c c c c c c c c c Y*Y*Y*Y* c c c]
v^1 = [c c Y* c c Y* c c c c c c Y* c c Y*]

At this point we need to symbolically deal with the $MixColumns(\mathbf{v}^1)$ output, as we will be working with the modified output that has been XOR'd with the constant E. As in [13], this is accomplished in practice by setting unknown constants c to zero, and calculating the output of the $MixColumns(\mathbf{v}^1)$ function. The unknown constants are all pushed into the variable E, which we never need to determine the true value of. This means our output of round $r = 1$ becomes:

m^1 = [Z*Z*Z*Z*Z*Z*Z*Z* c c c c Z*Z*Z*Z*]

Note that 4 bytes of this output are constant. We again set these constant bytes to zero to simplify our further manipulation of them. This means our input to the next round becomes:

p^2 = [Z*Z*Z*Z*Z*Z*Z*Z* 0 0 0 0 Z*Z*Z*Z*]

We are not able to recover $n_8^2, \cdots, n_{11}^2$ yet, as the inputs associated with those key bytes are constant.

We first attempt to recover n_0^2, which is performed for all 256 candidates for $k_9^1, \cdots, k_{12}^1$. As mentioned in Sect. 2.4, the highest correlation peak determines both $k_9^1, \cdots, k_{12}^1$ and n_0^2. This means we no longer have a group of candidates for the input, but a single value:

p^2= [Z Z Z Z Z Z Z Z 0 0 0 0 Z Z Z Z]

We can then proceed with the CPA attack on the remaining bytes of $\mathbf{n}^2$. Bytes $n_1^2, \cdots, n_6^2$ can be recovered by application of the CPA attack from Sect. 2.3.

Recovery of n_7^2 using the same process is not possible, as $MixColumns(\mathbf{v^1})$ interacts with the leakage model. The inputs to this round p_6^2 and p_7^2, are generated by the previous-round $MixColumns(\mathbf{v^1})$ outputs m_6^1 and m_7^1.

When attacking n_7^2, we apply (1) to (6) and (7). This means our leakage is:

$$HW\left((n_6^2 \oplus (6)) \oplus (n_7^2 \oplus (7))\right) \tag{9}$$

The XOR cancels common terms in (6) and (7), and in this case that cancels term s_1. As s_1 is the variable and known input to the $MixColumns(\mathbf{v^1})$, the leakage appears constant and the attack fails. Instead, we can recover this value using a CPA attack on the next round, which is described later.

Returning to our CPA attack on the modified round key, we are unable to recover $n_8^2, \cdots, n_{11}^2$ as the associated inputs are constant. As n_{11}^2 is unknown, we cannot directly recover $n_{12}^2, \cdots, n_{15}^2$. Instead we again use the method of Sect. 2.4 to generate 256 candidates for $n_{12}^2, \cdots, n_{15}^2$.

At this point we assume the CPA attack has succeeded, meaning we have recovered the following bytes of the *modified* round key, where the final 4 bytes are partially known – we have 256 candidates for this group, as we know the relationship between each byte, but simply don't know the starting byte to define the group:

$\mathbf{n^2}$ = [N N N N N N c c c c c N*N*N*N*]

Remember, once we apply $AddRoundKey(\mathbf{n^2}, \mathbf{p^2})$, the constant E will be removed – E is included in both the output of $MixColumns(\mathbf{v^1})$ and the modified key – meaning we can determine the true value of the input to $SubBytes()$.

The outputs $8, \cdots, 11$ of $MixColumns(\mathbf{v^1})$ from the first round are constant, so we also know the four unknown modified bytes $n_8^2, \cdots, n_{11}^2$ can be ignored at this point. The result of $AddRoundKey(\mathbf{n^2}, \mathbf{p^2})$ for these bytes will be another constant.

The unknown byte n_7^2 is associated with variable input data, meaning this output will be unknown and variable, which cannot be ignored. At this point we can represent the known outputs of $SubBytes()$ and $ShiftRows()$:

$\mathbf{s^2}$ = [Y Y Y Y Y Y Y ? c c c c Y* Y*Y*Y*]

$\mathbf{v^2}$ = [Y Y c Y* Y c Y* Y c Y* Y ? Y* Y Y c]

As before, we can set unknown constant values to zero to determine the modified output $\mathbf{m^2} = MixColumns(\mathbf{v^2})$. The unknown variable byte means 4 bytes of the $MixColumns(\mathbf{v^2})$ output are currently unknown. In addition, we have 256 candidates for the remaining known values, since the four modified bytes $n_{12}^2, \cdots, n_{15}^2$ have been mixed into all output bytes by $ShiftRows(\mathbf{p^2})$ and $MixColumns(\mathbf{v^2})$:

$\mathbf{m^2}$ = [Z*Z*Z*Z*Z*Z*Z*Z* ? ? ? ? Z*Z*Z*Z]

This becomes the input to the next round:

$\mathbf{p^3}$ = [Z*Z*Z*Z*Z*Z*Z*Z* ? ? ? ? Z*Z*Z*Z]

We again apply the CPA attack on n_0^3 across all values for n_0^3 and the 256 candidates for the previous modified round key (a total of 2^{16} guesses), the peak telling us the value of n_0^3 and $n_{12}^2, \cdots, n_{15}^2$. We now know which of the candidates to select for further processing:

$\mathbf{p}^3 =$ [Z Z Z Z Z Z Z Z ? ? ? ? Z Z Z Z]

We can apply a CPA attack to discover the modified key values $n_1^3, \cdots, n_7^3$. The unknown plaintext byte ? represents a changing value. We cannot ignore it as we can constant values in the $MixColumns(\mathbf{v}^2)$, and thus cannot apply the CPA attack on the remaining bytes.

Instead we enumerate all possibilities for n_7^2, and apply a CPA attack against n_8^3, similarly to previously described attacks from Sect. 2.4. We verified experimentally that the correlation value with the highest peak for n_8^3 resulted only when n_7^2 was the correct value. This means we now have the entire modified output of $MixColumns(\mathbf{v}^2)$, and thus the complete modified input plaintext to round 3:

$\mathbf{p}^3 =$ [Z Z Z Z Z Z Z Z Z Z Z Z Z Z Z Z]

With n_7^2 and n_8^3 now known, we can continue with the CPA attack against $n_9^3, \cdots, n_{15}^3$. At this point we have an entire modified key:

$\mathbf{n}^3 =$ [N N N N N N N N N N N N N N N N]

We can again apply the modified key $\mathbf{n}^3$ to the modified output of the previous round $\mathbf{m}^2$ to recover the complete output of round $r = 3$, which will be the actual input to round $r = 4$. This allows us to perform a CPA attack and recover the true round key $\mathbf{k}^4$. This round key can then be rolled backwards using the AES key schedule to determine the original encryption key.

We have now attacked an AES-CCM* implementation as specified in the IEEE 802.15.4 standard. This attack requires only the control of the four bytes of `FrameCounter`, which are sent as plaintext over the air.

The computational load of the attack is minimal: performing these steps on an Intel i5-2540M laptop using a single thread program written in C++ takes under ten minutes with 20 000 traces, using only the subset of points in each trace from Table 1. Note when performing the hypothetical value calculation for intermediate rounds, the calculation was accelerated using the Intel AES-NI instruction set for performing the $SubBytes()$, $ShiftRows()$, and $MixColumns()$ operations, which form part of a single AES round executed by this instruction [14].

5   Attacking Wireless Nodes

In the previous sections, we demonstrated the vulnerability of an IEEE 802.15.4 SoC device to power analysis, and how the AES-CCM* mode used during reception of an encrypted IEEE 802.15.4 packet can be attacked when the underlying hardware is vulnerable to power analysis. The last two aspects of this attack are to (1) demonstrate how we can trigger that encryption operation, and (2) determine where in the power signature the encryption occurred.

Details of the required packet format for reception are detailed in the extended version of this paper. The packet must simply conform to IEEE 802.15.4 requirements and have valid addressing information. The attacker controls the `FrameCounter` field as part of the attack.

In order for the side-channel attack to be successful, the attacker needs to determine *when* the AES encryption is occurring. As a starting point, the attacker can use information on when the frame should have been received by the target node. Practically, this would be either the attacker's transmitter node toggling an IO line when the packet goes over the air, or the attacker could use another node that also receives the transmitted messages to toggle an IO line.

To determine the reliability of such a trigger, we measured the time between the frame being received and the actual start of AES encryption on the target node. Over 100 transmitted frames the delay varied between 311 and 338 μs. The mean value of the delay was 325 μs (5200 clock cycles), with a standard deviation of 7 μs (112 clock cycles). The jitter in the delay is assumed to be from the software architecture, which uses an event queue process the frames. Solutions for aligning or resynchronizing power traces before applying power analysis is well known [15–18].

To test the ability of an attacker to realign captured power traces, we used a simple normalized cross-correlation algorithm [19] to match a feature across multiple power traces for realignment, performing a simple static alignment [20].

The selected feature was a window at 9.2–29.2 μs after the start of the AES encryption in one reference trace, meaning the matched feature extended slightly beyond the actual AES encryption. We confirmed that a high correlation peak was generated only for a single sample around the AES algorithm with many sample power traces. A threshold of 0.965 on the correlation output (determined empirically) was used; if a power trace had no correlation peak higher than this level, the trace was dropped.

Future work on this IEEE 802.15.4 attack can include applying more advanced preprocessing techniques (such as differential frequency analysis or principal component analysis). But such preprocessing techniques are not required to fundamentally prove that (a) the AES core is leaking, and (b) the AES operation has some unique signature allowing realignment to succeed.

6   Conclusions

The IEEE 802.15.4 wireless standard is a popular lower layer for many protocols being used in or marketed for the coming "Internet of Things" (see Sect. 1 for an enumeration of some of these). Such protocols often use the same underlying AES primitive as the IEEE 802.15.4 layer for security purposes.

This paper has demonstrated vulnerabilities in a real IEEE 802.15.4 wireless node. A successful attack against the AES peripheral present in the ATMega128RFA1 device was demonstrated. This attack was demonstrated against AES-ECB; as electronic code book (ECB) is not the operating mode of AES used in the network, we extended a previous attack on AES-CTR mode [13] to work against the AES-CCM* mode used in IEEE 802.15.4. This demonstrated that it is possible to recover the encryption key of a wireless node using side-channel power attacks and valid IEEE 802.15.4 messages sent to the node.

An extended version of this conference paper with additional details of the attack is available at https://eprint.iacr.org/2015/529.

Acknowledgments. The authors would like to thank the anonymous reviewers at COSADE 2016 for their insightful comments. Colin O'Flynn is funded by the Natural Sciences and Engineering Research Council of Canada (NSERC) under the CGS program.

References

1. IEEE: Standard 802.15.4-2006: Wireless Medium Access Control (MAC) and Physical Layer (PHY) Specifications for Low-Rate Wireless Personal Area Networks (WPANs) (2006)
2. Kocher, P.C., Jaffe, J., Jun, B.: Differential power analysis. In: Wiener, M. (ed.) CRYPTO 1999. LNCS, vol. 1666, pp. 388–397. Springer, Heidelberg (1999)
3. Brier, E., Clavier, C., Olivier, F.: Correlation power analysis with a leakage model. In: Joye, M., Quisquater, J.-J. (eds.) CHES 2004. LNCS, vol. 3156, pp. 16–29. Springer, Heidelberg (2004)
4. Gandolfi, K., Mourtel, C., Olivier, F.: Electromagnetic analysis: concrete results. In: Koç, Ç.K., Naccache, D., Paar, C. (eds.) CHES 2001. LNCS, vol. 2162, pp. 251–261. Springer, Heidelberg (2001)
5. Agrawal, D., Rao, J.R., Rohatgi, P.: Multi-channel attacks. In: Walter, C.D., Koç, Ç.K., Paar, C. (eds.) CHES 2003. LNCS, vol. 2779, pp. 2–16. Springer, Heidelberg (2003)
6. O'Flynn, C., Chen, Z.: A case study of side-channel analysis using decoupling capacitor power measurement with the OpenADC. In: Garcia-Alfaro, J., Cuppens, F., Cuppens-Boulahia, N., Miri, A., Tawbi, N. (eds.) FPS 2012. LNCS, vol. 7743, pp. 341–356. Springer, Heidelberg (2013)
7. de Meulenaer, G., Standaert, F.-X.: Stealthy compromise of wireless sensor nodes with power analysis attacks. In: Chatzimisios, P., Verikoukis, C., Santamaría, I., Laddomada, M., Hoffmann, O. (eds.) MOBILIGHT 2010. LNICST, vol. 45, pp. 229–242. Springer, Heidelberg (2010)
8. Atmel Corporation: ATmega128RFA1 Datasheet (2014)
9. Kizhvatov, I.: Side channel analysis of AVR XMEGA crypto engine. In: Proceedings of the 4th Workshop on Embedded Systems Security, WESS 2009, pp. 8:1–8:7. ACM, New York (2009)
10. O'Flynn, C., Chen, Z.D.: ChipWhisperer: an open-source platform for hardware embedded security research. In: Prouff, E. (ed.) COSADE 2014. LNCS, vol. 8622, pp. 243–260. Springer, Heidelberg (2014)
11. Standaert, F.-X., Malkin, T.G., Yung, M.: A unified framework for the analysis of side-channel key recovery attacks. In: Joux, A. (ed.) EUROCRYPT 2009. LNCS, vol. 5479, pp. 443–461. Springer, Heidelberg (2009)
12. Whiting, D., Ferguson, N., Housley, R.: Counter with CBC-MAC (CCM). https://tools.ietf.org/html/rfc3610

13. Jaffe, J.: A first-order DPA attack against AES in counter mode with unknown initial counter. In: Paillier, P., Verbauwhede, I. (eds.) CHES 2007. LNCS, vol. 4727, pp. 1–13. Springer, Heidelberg (2007)
14. Gueron, S.: Intel Advanced Encryption Standard (AES) new instructions set. Whitepaper Doc. No. 323641-001 (2012)
15. Clavier, C., Coron, J.-S., Dabbous, N.: Differential power analysis in the presence of hardware countermeasures. In: Paar, C., Koç, Ç.K. (eds.) CHES 2000. LNCS, vol. 1965, pp. 252–263. Springer, Heidelberg (2000)
16. Gebotys, C.H., Ho, S., Tiu, C.C.: EM analysis of Rijndael and ECC on a wireless Java-based PDA. In: Rao, J.R., Sunar, B. (eds.) CHES 2005. LNCS, vol. 3659, pp. 250–264. Springer, Heidelberg (2005)
17. van Woudenberg, J.G.J., Witteman, M.F., Bakker, B.: Improving differential power analysis by elastic alignment. In: Kiayias, A. (ed.) CT-RSA 2011. LNCS, vol. 6558, pp. 104–119. Springer, Heidelberg (2011)
18. Batina, L., Hogenboom, J., van Woudenberg, J.G.J.: Getting more from PCA: first results of using principal component analysis for extensive power analysis. In: Dunkelman, O. (ed.) CT-RSA 2012. LNCS, vol. 7178, pp. 383–397. Springer, Heidelberg (2012)
19. Lewis, J.P.: Fast template matching. In: Canadian Conference on Vision Interface – VI 1995, pp. 120–123 (1995)
20. Mangard, S., Oswald, E., Popp, T.: Power Analysis Attacks: Revealing the Secrets of Smart Cards. Springer, New York (2007)

Improved Side-Channel Analysis Attacks on Xilinx Bitstream Encryption of 5, 6, and 7 Series

Amir Moradi[✉] and Tobias Schneider

Horst Görtz Institute for IT Security, Ruhr-Universität Bochum, Bochum, Germany
{amir.moradi,tobias.schneider-a7a}@rub.de

Abstract. Since 2012, it is publicly known that the bitstream encryption feature of modern Xilinx FPGAs can be broken by side-channel analysis. Presented at CT-RSA 2012, using graphics processing units (GPUs) the authors demonstrated power analysis attacks mounted on side-channel evaluation boards optimized for power measurements. In this work, we extend such attacks by moving to the EM side channel to examine their practical relevance in real-world scenarios. Furthermore, by following a certain measurement procedure we reduce the search space of each part of the attack from 2^{32} to 2^8, which allows mounting the attacks on ordinary workstations. Several Xilinx FPGAs from different families – including the 7 series devices – are susceptible to the attacks presented here.

1 Introduction

Side-Channel Analysis (SCA) attacks have become a serious threat to cryptographic implementations. This indeed has been highlighted by publicly reporting several successful attacks on commercial devices, e.g., [1,5,9,14,15,20]. One of the well-known examples are the attacks on the bitstream encryption feature of FPGA devices which also garnered the attention of (industry and academic) FPGA communities.

The first SCA attack on the bitstream encryption of (out-dated and discontinued) Xilinx Virtex-II pro family has been presented in [11], where a full 168-bit key of the underlying triple-DES algorithm could be recovered by a single power-up of the FPGA ($\approx$70,000 traces) by searching in a space of 2^6 for each 6-bit part of the key. The second work [12] showed that a similar attack on more recent Xilinx FPGA families (Virtex-4 and Virtex-5) is feasible. However, due to the underlying AES-256 algorithm and the implementation architecture, the presented attack could only recover the key by searching in a space of 2^{32} for each 32-bit part of the key. To deal with such a complexity, the authors made use of four graphics processing units (GPUs with a total of 4×448 thread processors) and mounted the attack on a single point of the 60,000 power traces collected from a single power-up of the FPGA. The full 256-bit key could be recovered in 4.5 h by such a setup while the attack on the second round

© Springer International Publishing Switzerland 2016
F.-X. Standaert and E. Oswald (Eds.): COSADE 2016, LNCS 9689, pp. 71–87, 2016.
DOI: 10.1007/978-3-319-43283-0_5

(to recover the second 128-bit key) was not as efficient as that on the first round. In all the aforementioned attacks, power traces of various SASEBO or SAKURA boards have been collected. Since such boards are explicitly designed for power analysis evaluation purposes, remounting the same attacks on real-work applications might be challenging, where PCB should be slightly modified to provide a suitable measurement point.

As a side note, similar attacks on Altera FPGAs (Stratix-II and Stratix-II families) have been later reported in [13, 17]. Compared to that on Xilinx FPGAs, the attacks required a reverse-engineering step (of the software development tools) and a sophisticated measurement procedure to deal with the underlying AES algorithm in counter mode.

Our Contribution. In this work we present an improved attack on bitstream encryption of modern Xilinx FPGAs. Our achievements can be summarized as follows:

- By further investigation of the design architecture of the AES decryption module, we present a more suitable power model for the attacks, particularly on the second cipher round.
- By means of a dedicated measurement setup, we reduce the search space from 2^{32} for each part of the attack to 2^8. Therefore, the attacks can be performed using ordinary desktop computers.
- We present the result of the attacks on Virtex-5, Spartan-6, Kintex-7, and Artix-7 FPGAs as the samples of 5, 6, and 7 series.
- In contrast to all reported attacks on Xilinx bitstream encryption, we present the results via electro magnetic (EM) side channel.

In short, we avoid the need of using GPUs, and demonstrate strong and efficient attacks on bitstream encryption of 7 series FPGAs of Xilinx which are currently in production.

2 Preliminaries

2.1 Xilinx Bitstream Encryption

Bitstream encryption, in general, has been introduced to prevent cloning and counterfeiting the user designs. In order to protect proprietary algorithms, secret materials, and obfuscated designs from reverse engineering, it is essential for the user to employ bitstream encryption. Xilinx products are mainly SRAM-based FPGAs, which implies reconfiguration (loading bitstream into the FPGA) every time the FPGA powers up. Since the bitstream has to be stored outside the FPGA (in a non-volatile memory), bitstream encryption is a must-to-have feature for the FPGA vendors, whose products are based on volatile memory (e.g., Xilinx).

The current available FPGA series of Xilinx make use of AES-256 in cipher block chaining (CBC) mode to encrypt the bitstream. Suppose that the bitstream is divided into n 128-bit blocks $p^{i \in \{1,\dots,n\}}$. The encrypted bitstream, which is formed by n 128-bit blocks c^i, is generated by

$$c^i = \text{AES}_k^{\text{ENC}}(p^i \oplus c^{i-1}),$$

assuming $c^0 = IV$. The secret key k and the initialization vector IV can be arbitrary selected by the user. The Xilinx development tools generate a human-readable ASCI file (with .nky extension) of the selected key and IV, which is given to the programming device to store the key inside the FPGA. As a side note, although IV is written into the .nky file, the programming device stores only the key into the FPGA via the JTAG port. Older versions of the Xilinx FPGAs make use of only volatile memory for the key storage which requires an external battery during power shortage. The newer families are, in addition, equipped with one-time programmable fuses.

Although there are not many public documents about the details of the structure of the bitstream file, with moderate efforts (similar to that of [11,12]) the essential information can be revealed (e.g., bit and byte endianness and the size of the header before the encrypted part starts). Such an investigation recovered that IV (in plain) is available in the bitstream before the encrypted part starts. Further, this IV must not be necessary the same as the one which has been formerly written into the .nky file.

2.2 Configuration and Measurement

The encrypted bitstream can be sent to the FPGA via several different protocols (serial, parallel, master, slave, and JTAG). Since the JTAG port is dedicated to configuration (and it has to be used for key programming), such a port is usually available in most of the real-world applications (e.g., set up boxes). In [12], a customized micro-controller (MCU) has been used to configure the FPGA (via JTAG) and provide a trigger signal for the oscilloscope. In [11,12], by monitoring the voltage drop of a resistor in VDD$_{\text{int}}$ path, power consumption traces of the FPGA have been collected. The decryption module inside the FPGA receives 128-bit ciphertext blocks $c^{i \in \{1,\dots,n\}}$ in a consecutive fashion, and derives the plaintexts p^i as

$$p^i = \text{AES}_k^{\text{DEC}}(c^i) \oplus c^{i-1},$$

with $c^0 = IV$.

It has been reported in [12] that – in addition to the decryption engine – other modules of the FPGA are active whose energy consumption (as noise) are visible through the measured power traces. Hence, filtering the traces to reduce the noise was essential. As shown in Fig. 1, the decryption of c^i takes place when the next block c^{i+1} is fed into the FPGA. Further, the decryption clock is somehow synchronized with the JTAG clock.

2.3 Attack

Since AES-256 consists of 14 rounds, which fits to the 14 visible peaks in the power trace, it has been assumed that the decryption module in the FPGA realizes a round-based architecture of the AES-256, which performs one cipher round at each single clock cycle [12]. Figure 2 shows the hypothetical design architecture that has been considered in [12].

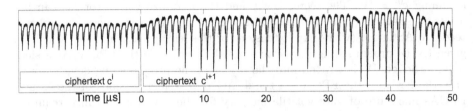

Fig. 1. A sample power trace of Spartan-6 (with 20 MHz low-pass filter) during loading an encrypted bitstream

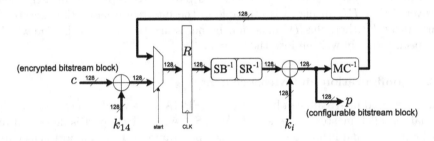

Fig. 2. Hypothetical design architecture of the AES-256 decryption module of modern Xilinx FPGAs (taken from [12])

Assuming such an architecture, the state register R stores $R_1 = c \oplus k_{14}$ and $R_2 = \mathrm{MC}^{-1}\Big(\mathrm{SR}^{-1}\big(\mathrm{SB}^{-1}\left(c \oplus k_{14}\right)\big)\Big) \oplus k_{13}$ at the first and second cipher rounds respectively[1]. In general, at round $1 < i < 15$ the content of the state register is $R_i = \mathrm{MC}^{-1}\Big(\mathrm{SR}^{-1}\big(\mathrm{SB}^{-1}\left(R_{i-1}\right)\big)\Big) \oplus k_{15-i}$. Indeed, the above shown hypothetical architecture has been verified by examining the correlation between the measured power traces and the Hamming distance (HD) of the state register in a known-key settings. As shown in Fig. 3, the power traces show a clear dependency on $\mathrm{HD}(R_1, R_2)$. However, such a dependency is strongly mitigated (but still available) in the next cipher round, i.e., on $\mathrm{HD}(R_2, R_3)$.

Let us denote $R_1 \oplus R_2$ by Δ_{R_1,R_2} and its byte at position $i \in \{0,\dots,15\}$ with $\Delta^{(i)}_{R_1,R_2}$. Following the AES notations, we represent the first column of the state R by $R^{(0,1,2,3)}$. Hence, due to the linear property of the MixColumns and

---

[1] MC: MixColumns, SR: ShiftRows, SB: SubBytes.

its inverse we can write

$$\Delta_{R_1,R_2}^{(0,1,2,3)} = R_1^{(0,1,2,3)} \oplus R_2^{(0,1,2,3)}$$

$$= c^{(0,1,2,3)} \oplus k_{14}^{(0,1,2,3)} \oplus \text{M1C}^{-1}\Big(\Big\langle S^{-1}\Big(c^{(0)} \oplus k_{14}^{(0)}\Big), S^{-1}\Big(c^{(13)} \oplus k_{14}^{(13)}\Big),$$

$$S^{-1}\Big(c^{(10)} \oplus k_{14}^{(10)}\Big), S^{-1}\Big(c^{(7)} \oplus k_{14}^{(7)}\Big)\Big\rangle\Big) \oplus \text{M1C}^{-1}\Big(k_{13}^{(0,1,2,3)}\Big), \qquad (1)$$

where S^{-1} stands for the Sbox inverse, and M1C^{-1} for the inverse of the Mix-Columns operation on a single column.

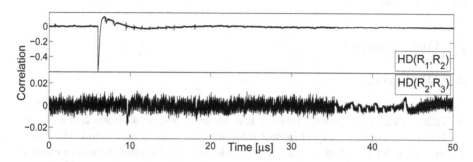

Fig. 3. Spartan-6, correlation between **power** traces and $\text{HD}(R_1, R_2)$ and $\text{HD}(R_2, R_3)$

Since both $k_{14}^{(0,1,2,3)}$ and $\text{M1C}^{-1}\Big(k_{13}^{(0,1,2,3)}\Big)$ are fixed and independent of the ciphertext c, correlation power analysis (CPA) [2] (respectively classical DPA [7]) attacks, that target bits of $\Delta_{R_1,R_2}^{(0,1,2,3)}$, can be performed by guessing four key bytes $\Big\langle k_{14}^{(0)}, k_{14}^{(13)}, k_{14}^{(10)}, k_{14}^{(7)}\Big\rangle$. Such a 2^{32}-bit attack (on a single point of the power traces) has been performed in [12] using GPUs. The same attack with the same principle can be performed on the other columns of the Δ_{R_1,R_2} to recover full 128-bit round key k_{14}.

Having k_{14}, we can follow the same procedure for the second cipher round. Let us denote $\text{MC}^{-1}\Big(\text{SR}^{-1}(\text{SB}^{-1}(c \oplus k_{14}))\Big)$ by c' and $\text{MC}^{-1}(k_{13})$ by k_{13}'. As an example, for the first column of Δ_{R_2,R_3} we can write

$$\Delta_{R_2,R_3}^{(0,1,2,3)} = R_2^{(0,1,2,3)} \oplus R_3^{(0,1,2,3)}$$

$$= c'^{(0,1,2,3)} \oplus k_{13}'^{(0,1,2,3)} \oplus \text{M1C}^{-1}\Big(\Big\langle S^{-1}\Big(c'^{(0)} \oplus k_{13}'^{(0)}\Big), S^{-1}\Big(c'^{(13)} \oplus k_{13}'^{(13)}\Big),$$

$$S^{-1}\Big(c'^{(10)} \oplus k_{13}'^{(10)}\Big), S^{-1}\Big(c'^{(7)} \oplus k_{13}'^{(7)}\Big)\Big\rangle\Big) \oplus \text{M1C}^{-1}\Big(k_{12}^{(0,1,2,3)}\Big). \qquad (2)$$

The same attacks (as shown in [12]) can target the bits of $\Delta_{R_2,R_3}^{(0,1,2,3)}$ and search in a space of 2^{32} to recover $\Big\langle k_{13}'^{(0)}, k_{13}'^{(13)}, k_{13}'^{(10)}, k_{13}'^{(7)}\Big\rangle$. The same procedure is repeated for other columns of Δ_{R_2,R_3}, and after revealing k_{14} and k_{13}' the 256-bit main key can be derived.

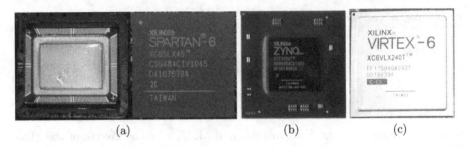

Fig. 4. Different packaging technologies: (a) wire-bond, (b) flip-chip, (c) flip-chip with lid-heat spreader

3   Our Analysis

3.1   Packaging

In contrast to all the reported SCA attacks mounted on bitstream encryption of Xilinx devices, we concentrate on EM analysis. Figure 4 shows two different packaging technologies flip-chip and wire-bond. In case of wire-bond, the metal layers (of the FPGA chip) are at the top side, and the bonding wires are covered by molding components (usually plastic, see Fig. 4(a)). Since the main EM radiations are due to the current flowing through VDD path(s), the EM probes can be placed at the top of the chip, if the top metal layers include the VDD_{int} (see Fig. 5(b)). For the flip-chip technology, sometimes the top of the chip is covered by a lid-heat spreader (Fig. 4(c)), which must be removed for EM analysis. Compared to the wire-bond case, the silicon side of the chip (usually a thick layer) is accessible, which prevents reaching the layers carrying VDD. Hence, the EM signals are usually weak unless the thick silicon is thinned by means of sophisticated polishing devices, that also allows using localized EM microprobes [6].

3.2   Measurements

For the EM measurements we used a digital oscilloscope at a sampling rate of 5 GS/s and bandwidth of 1.5 GHz. We have employed only near-field probes of LANGER EMV-Technik. Further, depending on the amplitude of the signal, we made use of one or two high-bandwidth AC amplifiers ZFL-1000LN+ from Mini-Circuits.

Depending on the packaging, type of the FPGA, and the visibility of the signal, we used either RF-U5-2 or RF-R50-1 EM probes. In case of Virtex-5 and Kintex-7 (both with flip-chip) as well as Artix-7 (wire-bond) we achieved the best results with a RF-R50-1 probe, and for Spartan-6 (wire-bond) with a RF-U5-2 probe (see Fig. 5). Except removing the lid-heat spreader of the Virtex-5 FPGA, we did not modify the packaging of the FPGAs.

For the sake of simplicity, we concentrate on the Spartan-6 case, and discuss the other FPGAs at the end of this section. We also developed a MCU-based device to configure the FPGAs through the JTAG port. Figure 6 shows a single EM trace of the Spartan-6 FPGA (synchronized with that of Fig. 1). As a proof of concept, and to verify the hypothetical design architecture, in a known-key scenario we measured 100,000 traces and estimated the correlation considering $HD(R_i, R_{i+1})$, $0 < i < 14$. The results, which are shown in Figure 6, indicate that the high correlation only exist at the first cipher round, which makes the attacks challenging at the second round.

(a) Virtex-5   (b) Spartan-6   (c) Kintex-7   (d) Artix-7

Fig. 5. EM probes and different FPGAs, (a) XC5VLX50-1FFG324, (b) XC6SLX75-2CSG484C, (c) XC7K160T-1FBGC, (d) XC7A35T-1CPG236C

We have tried many different hypotheses for the design architecture, and finally the highest correlation has been observed considering the same architecture as shown in Fig. 2 but with $HD(R_1, R_{i+1})$, $0 < i < 14$ model. Although no design architecture can justify why the SCA leakage depends on the state register at round $i+1$ and that of the first round, such a model leads to considerably high correlations[2] as shown in Fig. 6.

The previous attacks have been based on measuring one or multiple power-ups of the FPGA [12]. This means that the ciphertexts have been previously defined (stored in a non-volatile external memory). Instead, we aim at selecting the ciphertexts by our choice. Sending chosen cipherexts to the FPGA, however, has a negative consequence on the interconnections of the FPGA. The switch boxes and look-up tables are wrongly configured which leads to short circuits (high power consumption and high temperature) and may destruct certain modules. Therefore, in order to avoid such consequences, after sending one (or a couple of) chosen ciphertext(s), the configuration process should be restarted. This can be easily done by sending certain commands through the JTAG port, which are available in Xilinx public documents, e.g., [18]. Following such instructions, we adjusted our MCU-based programmer to perform a configuration reset after each single measurement. In more details, after starting the configuration process the MCU device sends the header (the unencrypted part of the bitstream), the chosen ciphertext, and a dummy 128-bit ciphertext block. When the dummy ciphertext is sent, the corresponding EM/power trace is measured,

---

[2] As a side note, we found this leakage model by coincidence, and it is valid for all considered FPGAs and for both power and EM leakages.

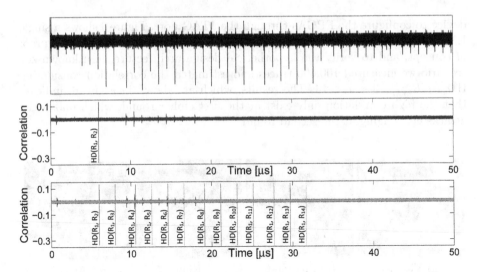

Fig. 6. Spartan-6, EM analysis, (top) a sample trace, (middle) correlation between EM traces and $HD(R_i, R_{i+1})$ and (bottom) $HD(R_1, R_{i+1})$, $0 < i < 14$

since – as stated in Sect. 2 and shown in Fig. 1 – the decryption of the first ciphertext takes place when the second ciphertext block is sent.

3.3   Attacks

As explained in Sect. 2, the previous attack needs to search in a space of at least 2^{32}. Recalling Eq. (1), if ciphertext bytes $c^{(13)}$, $c^{(10)}$, and $c^{(7)}$ are constant we can write

$$
\begin{aligned}
\Delta_{R_1,R_2}^{(0,1,2,3)} =\; & c^{(0,1,2,3)} \oplus k_{14}^{(0,1,2,3)} \oplus \text{M1C}^{-1}\!\left(\!\left\langle S^{-1}\!\left(c^{(0)} \oplus k_{14}^{(0)}\right), S^{-1}\!\left(c^{(13)} \oplus k_{14}^{(13)}\right),\right.\right. \\
& \left.\left. S^{-1}\!\left(c^{(10)} \oplus k_{14}^{(10)}\right), S^{-1}\!\left(c^{(7)} \oplus k_{14}^{(7)}\right)\right\rangle\right) \oplus \text{M1C}^{-1}\!\left(k_{13}^{(0,1,2,3)}\right) \\
=\; & \left\langle \{0e\} \bullet S^{-1}\!\left(c^{(0)} \oplus k_{14}^{(0)}\right) \oplus c^{(0)} \oplus \delta^{(0)},\right. \\
& \{09\} \bullet S^{-1}\!\left(c^{(0)} \oplus k_{14}^{(0)}\right) \oplus c^{(1)} \oplus \delta^{(1)}, \\
& \{0d\} \bullet S^{-1}\!\left(c^{(0)} \oplus k_{14}^{(0)}\right) \oplus c^{(2)} \oplus \delta^{(2)}, \\
& \left. \{0b\} \bullet S^{-1}\!\left(c^{(0)} \oplus k_{14}^{(0)}\right) \oplus c^{(3)} \oplus \delta^{(3)}\right\rangle \\
=\; & \Delta_{R_1,R_2}^{\prime(0,1,2,3)} \oplus \delta^{(0,1,2,3)}, \tag{3}
\end{aligned}
$$

where constants $\{0e\}, \dots, \{0b\}$ are with respect to the MixColumns Inverse operation, and $\bullet$ the multiplication in $GF(2^8)$. Further, $\delta^{(0)}, \dots, \delta^{(3)}$ represent constants that depend on key k and ciphertext bytes 13, 10, and 7. If – in contrast to [12] – we select the ciphertexts which are given to the decryption module, and keep certain ciphertext bytes fixed (13, 10, and 7), we can perform CPA/DPA attacks by searching in a shorter spaces – as explained below – to find $k_{14}^{(0)}$.

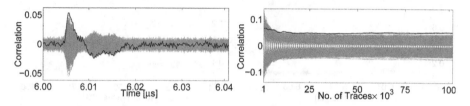

Fig. 7. Spartan-6, EM analysis, CPA in 2^{16}, $HD(\Delta^{(0)}_{R_1,R_2})$ model, (a) using 100,000 traces, (b) over the number of traces

Search in a Space of 2^{16}. For example, based on Eq. (3) $\Delta^{(0)}_{R_1,R_2}$ can be predicted by guessing $k^{(0)}_{14}$ and $\delta^{(0)}$, i.e., 16 bits. Therefore, $HD(\Delta^{(0)}_{R_1,R_2})$ can be used and a CPA can be performed accordingly. In this case, the disadvantage is the way that the constant $\delta^{(0)}$ contributes into the HD model. Since $\Delta^{(0)}_{R_1,R_2}$ and $\delta^{(0)}$ are linearly proportional, the use of HD model faces the *ghost peak* issue [10]. The result of such a 2^{16} attack with 100,000 traces as well as over the number of traces are shown in Fig. 7. Similarly, other bytes $\Delta^{(i\in\{1,2,3\})}_{R_1,R_2}$ can be predicted to find $k^{(0)}_{14}$ by searching in a 2^{16} space.

Search in a Space of 2^8. Similar to that of [12], the CPA/DPA attacks can be mounted targeting the bits of $\Delta'^{(0,1,2,3)}_{R_1,R_2}$ by guessing only 8-bit $k^{(0)}_{14}$ (see Eq. (3)). Due to the 32-bit size of $\Delta'^{(0,1,2,3)}_{R_1,R_2}$, 32 different attacks with the same target $k^{(0)}_{14}$ can be performed. Since predicting one single-bit flip out of a 128-bit register certainly leads to a low signal-to-noise ratio [10], it is favorable to combine the results of these 32 different attacks.

Heuristics. For a guessed key byte k, let us denote the result of the i-th CPA on sample point j by $\rho^{(i)}_{k,j}$. Following a similar approach to [3], we combine the results of multiple CPAs with different models by summing them up. As the constant is unknown we have to add the absolute values of the correlations as

$$\rho_{k,j} = \sum_{i=1}^{32} \left| \rho^{(i)}_{k,j} \right|$$

to combine the results of all 32 attacks. Figure 8 shows the corresponding results. It should be noted that – in contrast to their combination – none of the 32 single-bit CPA attacks could clearly distinguish the correct key byte $k^{(0)}_{14}$. Indeed, the complexity of the attack in this setting is 32×2^8.

Joint Probability. Let us suppose that the result of each CPA is a set of probabilities corresponding to the ranked key candidates. In other words, suppose that the i-th CPA on sample point j returns $p^{(i)}_{k,j}$ as the probability of the key byte k being the correct one. Since the 32 CPAs are independent of each other,

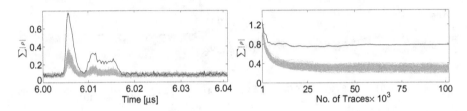

Fig. 8. Spartan-6, EM analysis, bitwise CPAs in 2^8, targeting bits of $\Delta'^{(0,1,2,3)}_{R_1,R_2}$, combined by absolute sum, (a) using 100,000 traces, (b) over the number of traces

we can combine the results as

$$p_{k,j} = \prod_{i=1}^{32} p_{k,j}^{(i)}. \tag{4}$$

At this step, the question raised is how to project the correlation values, i.e., the result of the CPA, to probabilities? Following the concept presented in [10] and also employed in [4], we can apply Fisher's z-transform and normalize the result as

$$r_{k,j}^{(i)} = \frac{1}{2\sqrt{N-3}} \ln \left(\frac{1 + \rho_{k,j}^{(i)}}{1 - \rho_{k,j}^{(i)}} \right),$$

where N is the number of traces used in the CPA. Now, $r_{k,j}^{(i)}$ is a sample that can be (approximately) interpreted according to the normal distribution $\mathcal{N}(0,1)$. Therefore, we can project it to probability by

$$p_{k,j}^{(i)} = 2 \int_{-\left| r_{k,j}^{(i)} \right|}^{0} \mathsf{PDF}_{\mathcal{N}(0,1)}(t)dt = 1 - 2\mathsf{CDF}_{\mathcal{N}(0,1)} \left(-\left| r_{k,j}^{(i)} \right| \right),$$

where $\mathsf{PDF}_{\mathcal{N}(0,1)}$ and $\mathsf{CDF}_{\mathcal{N}(0,1)}$ are respectively the probability density and cumulative distribution functions of the standard normal distribution.

We have followed this procedure and calculated the joint probabilities based on Eq. (4). The corresponding results, shown in Fig. 9, indicate that this scheme is also able to combine the results of all 32 CPAs and finally reveal the key. As a side note, the probabilities can also be combined following the Bayes' theorem. However, since the Bayes' theorem results in a set of probabilities with $\sum_{\forall k} p_{k,j}^{(i)} = 1$, for the sample points where none of the key candidates shows a high correlation, the probability of one key candidate (at that sample point) leads to a significantly higher value compared to that of the other candidates. This prevents us to find the most leaking sample points and distinguish the correct key. Hence, the corresponding results are omitted.

Linear Regression. From another perspective, we can map this problem to that which has been solved by means of linear regression (also known as

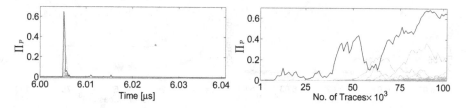

Fig. 9. Spartan-6, EM analysis, bitwise CPAs in 2^8, targeting bits of $\Delta'^{(0,1,2,3)}_{R_1,R_2}$, combined by joint probability, (a) using 100,000 traces, (b) over the number of traces

stochastic attacks) [16]. In other words, we suppose that by guessing $k^{(0)}_{14}$ the bits of $\Delta'^{(0,1,2,3)}_{R_1,R_2}$ contribute each with a certain weight to the leakage with respect to constants $\delta^{(0,1,2,3)}$. In more details, it is assumed that the leakage l at sample point j can be written as

$$l_j = \beta_{0,j} + \sum_{b=1}^{32} \beta_{b,j} g_b,$$

where g_b represents the b-th bit of $\Delta'^{(0,1,2,3)}_{R_1,R_2}$.

In order to find the coefficients $\beta_{b,j} \in \mathbb{R}$ – by following the procedure of [16] – for each guessed key, we form a matrix $\mathbf{M}$ as

$$\mathbf{M} = \begin{pmatrix} 1 & g_1^1 & g_2^1 & \cdot & \cdot & \cdot & g_{32}^1 \\ 1 & g_1^2 & g_2^2 & \cdot & \cdot & \cdot & g_{32}^2 \\ \cdot & \cdot & & & & & \cdot \\ \cdot & \cdot & & & & & \cdot \\ \cdot & \cdot & & & & & \cdot \\ 1 & g_1^N & g_2^N & \cdot & \cdot & \cdot & g_{32}^N \end{pmatrix},$$

where g_b^i represents the b-th bit of the predicted $\Delta'^{(0,1,2,3)}_{R_1,R_2}$ (based on the guessed $k^{(0)}_{14}$) for the i-th measurement (trace). As shown in [8], by means of the least square estimation, the vector of coefficients $\vec{\beta}_j = (\beta_{0,j}, \dots, \beta_{32,j})$ is estimated as

$$\vec{\beta}_j = \underbrace{\left(\mathbf{M}^T\mathbf{M}\right)^{-1}}_{\mathbf{A}} \underbrace{\mathbf{M}^T \vec{l}_j}_{\vec{\alpha}_j},$$

where $\mathbf{M}^T$ stands for the transpose of the matrix $\mathbf{M}$, and $\vec{l}_j$ for the vector of leakages at sample point j (i.e., N measured traces at sample point j). $\mathbf{A}$ is a matrix of 33×33 and independent of the sample points; hence, it can be derived with processing only the associated ciphertexts. The vector $\vec{\alpha}_j$ (formed by 33 elements) is also obtained for each sample point independently. Therefore, for each guessed $k^{(0)}_{14}$, all the measured traces are processed once to derive $\mathbf{A}$ and $\vec{\alpha}_j$, $\forall j$. Consequently, $\vec{\beta}_j$, $\forall j$ are derived by $\mathbf{A}^{-1}\vec{\alpha}_j$. At the next step, instead

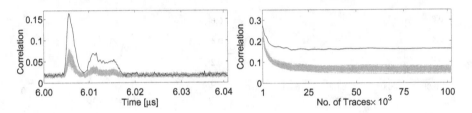

Fig. 10. Spartan-6, EM analysis, CPA in 2^8, weighted bits of $\Delta_{R_1,R_2}^{'(0,1,2,3)}$, recovered by linear regression, (a) using 100,000 traces, (b) over the number of traces

of the HD, the following model at sample point j for the i-th measurement is considered to perform a CPA:[3]

$$\hat{l}_j^i = \beta_{0,j} + \sum_{b=1}^{32} \beta_{b,j} g_b^i.$$

In other words, for each key hypothesis $k_{14}^{(0)}$, the measured traces are processed two times (first to derive the coefficients β and second to estimate the correlations). Figure 10 shows the results of this attack predicating that it outperforms all above shown attacks. Although it leads approximately to the same results as the *heuristic approach* (Fig. 8), its complexity is lower.

Other Key Bytes. The above explained procedure can be repeated for other key bytes. For example, by keeping the ciphertext bytes 0, 10, and 7 constant during the measurements, we can write

$$\Delta_{R_1,R_2}^{(0,1,2,3)} = \Big\langle \{0b\} \bullet S^{-1}\left(c^{(13)} \oplus k_{14}^{(13)}\right) \oplus c^{(0)} \oplus \delta^{(0)},$$
$$\{0e\} \bullet S^{-1}\left(c^{(13)} \oplus k_{14}^{(13)}\right) \oplus c^{(1)} \oplus \delta^{(1)},$$
$$\{09\} \bullet S^{-1}\left(c^{(13)} \oplus k_{14}^{(13)}\right) \oplus c^{(2)} \oplus \delta^{(2)},$$
$$\{0d\} \bullet S^{-1}\left(c^{(13)} \oplus k_{14}^{(13)}\right) \oplus c^{(3)} \oplus \delta^{(3)} \Big\rangle, \tag{5}$$

which allows the recovery of $k_{14}^{(13)}$.

It should be noted that the process of each column – in the first cipher round – is independent of the other columns. Hence, while all ciphertext bytes except 0, 4, 8, and 12 (the first row) are kept constant, four key recovery attacks (each by searching in a space of 2^8 to find the corresponding key bytes 0, 4, 8, and 12 of k_{14}) can independently be mounted. At the next step, a set of traces with fixed ciphertext bytes except the second row is measured which allows the recovery of

---

[3] We have also followed the suggestions of [8] to examine the squared error between the measured leakages l and estimated leakages $\hat{l}$, but our analyses showed better distinguishability when correlation is estimated instead.

key bytes 1, 5, 9, and 13. In short, we need to measure four sets of measurements, in each of which only the ciphertext bytes of one row are selected randomly, while the other 12 ciphertext bytes are kept constant (at any arbitrary value). With these four sets we are able to recover the full 128-bit last round key k_{14}.

Next Round. As the target algorithm is AES-256, we need to extend the attacks to the next decryption round. In contrast to that of [12], where Δ_{R_2,R_3} has been considered, based on our findings (presented in Sect. 3.2) we target Δ_{R_1,R_3}, i.e., the difference between the state register at the first and the third cipher rounds.

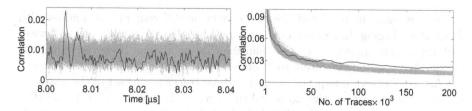

Fig. 11. Kintex-7, EM analysis, CPA in 2^8 second round, weighted bits of $\Delta_{R_1,R_2}^{\prime(0,1,2,3)}$, recovered by linear regression, (a) using 200,000 traces, (b) over the number of traces

Following the same principle as explained in Sect. 2.3 (particularly Eq. (2)) we can write

$$\Delta_{R_1,R_3}^{(0,1,2,3)} = R_1^{(0,1,2,3)} \oplus R_3^{(0,1,2,3)}$$

$$= c^{(0,1,2,3)} \oplus k_{14}^{(0,1,2,3)} \oplus \text{M1C}^{-1}\left(\left\langle S^{-1}\left(c'^{(0)} \oplus k_{13}'^{(0)}\right), S^{-1}\left(c'^{(13)} \oplus k_{13}'^{(13)}\right),\right.\right.$$

$$\left.\left. S^{-1}\left(c'^{(10)} \oplus k_{13}'^{(10)}\right), S^{-1}\left(c'^{(7)} \oplus k_{13}'^{(7)}\right)\right\rangle\right) \oplus \text{M1C}^{-1}\left(k_{12}^{(0,1,2,3)}\right), \qquad (6)$$

where $c' = \text{MC}^{-1}\left(\text{SR}^{-1}(\text{SB}^{-1}(c \oplus k_{14}))\right)$ and $k_{13}' = \text{MC}^{-1}(k_{13})$. By keeping $c'^{(13)}$, $c'^{(10)}$, and $c'^{(7)}$ constant, we can write

$$\Delta_{R_1,R_3}^{(0,1,2,3)} = \left\langle \{0e\} \bullet S^{-1}\left(c'^{(0)} \oplus k_{13}'^{(0)}\right) \oplus c^{(0)} \oplus \delta^{(0)},\right.$$

$$\{09\} \bullet S^{-1}\left(c'^{(0)} \oplus k_{13}'^{(0)}\right) \oplus c^{(1)} \oplus \delta^{(1)},$$

$$\{0d\} \bullet S^{-1}\left(c'^{(0)} \oplus k_{13}'^{(0)}\right) \oplus c^{(2)} \oplus \delta^{(2)},$$

$$\left. \{0b\} \bullet S^{-1}\left(c'^{(0)} \oplus k_{13}'^{(0)}\right) \oplus c^{(3)} \oplus \delta^{(3)}\right\rangle = \Delta_{R_1,R_3}^{\prime(0,1,2,3)} \oplus \delta^{(0,1,2,3)}.$$

$$(7)$$

Since the last round key k_{14} has been recovered, c' bytes can be arbitrary selected and the corresponding ciphertext $c = \text{SB}\left(\text{SR}(\text{MC}(c'))\right) \oplus k_{14}$ can be derived to be sent to the FPGA. Therefore, we followed the same procedure as explained for the first decryption round, and collected four sets of measurements, in each only one row of c' is selected randomly and the other bytes kept constant. This allows us to perform exactly the same attacks (each with complexity of 2^8) to find k'_{13} byte by byte to finally reveal the full 256-bit key. It is noteworthy that – in contrast to that of [12] – the attacks on the second round are as efficient as that on the first round since we are targeting Δ'_{R_1,R_3} instead of Δ'_{R_2,R_3}. As an example, the results of the attack on $k_{13}'^{(0)}$ (on the Kintex-7 device) are shown by Fig. 11.

3.4 Comparisons

Table 1 presents the results of the EM attacks on different FPGA families with different packaging. In general it can be concluded that the attacks on the devices with flip-chip technology is harder than the wire-bond ones. However, by shrinking the technology (from 65 nm to 28 nm) the attacks become harder as in case of the Artix-7 FPGA we required around 200,000 traces as one set of the measurements (with a constant row) to reveal the secrets[4], i.e., in total $2 \times 4 \times 200,000$ (1.6 million) traces. Since the FPGA device is in hand and control of the adversary, collecting more traces with chosen ciphertexts (if required) does not face a serious challenge. For example, with our setup we could collect each 100,000 traces of the chosen chiphertexts in around 90 min, which means that all 1.6 million traces[5] could be measured in less than a day.

Table 1. The attack performances

Family	5	6	7	
FPGA	Virtex-5	Spartan-6	Kintex-7	Artix-7
Package	Flip-chip	Wire-bond	Flip-chip	Wire-bond
Technology	65 nm	45 nm	28 nm	28 nm
Probe	RF-R50-1	RF-U5-2	RF-R50-1	RF-R50-1
Required traces for each set (row)	40,000	2,000	120,000	200,000

For the analyses, as shown in the attack results (Figs. 10 and 11) we have considered 200 sample points (either for the first or the second decryption round). These 200 sample points have been selected around the corresponding clock cycle based on the knowledge obtain from Fig. 6. We split each attack into two parts. The first part, which derives the matrix $\mathbf{A}$ and vectors $\vec{\alpha}_{j=1,...,200}$, for all 2^8

---

[4] We realized that other components on the PCB (BASYS 3 from www.digilentinc. com) introduce noise into the EM measurements.

[5] It is done in two parts since the second part can be started when k_{14} has already been recovered.

key candidates takes 21 min using an 8-core machine @3 GHz on 100,000 traces. The results are applied in the next corresponding key-recovery CPA on the same 100,000 traces, which also takes 12 min on the same machine. In total, for a full recovery (on both rounds) using in total 1.6 million traces we require $2 \times 16 \times 2 \times (21 + 12)$ min (around 1.5 days) using the aforementioned processing unit. These numbers for sure can be decreased by more parallelization or by reducing the number of considered sample points. It should be noted that since the attacks on Spartan-6 require far less number of traces, the measurements and analyses can be done in significantly shorter time, e.g., less than an hours for the measurements and the evaluations when each set of measurements contains only 2,000 traces.

3.5   Authentication

In Virtex-6 and 7 series devices, the bitstream encryption is integrated with an on-chip bitstream keyed-Hash Message Authentication Code (HMAC). It aims at authentication of the decrypted bitstream to prove that not even a single bit was modified. Stated in [19] "Without knowledge of the AES and HMAC keys, the bitstream cannot be loaded, modified, intercepted, or cloned. HMAC provides assurance that the bitstream provided for the configuration of the FPGA was the unmodified bitstream allowed to load".

As it is also mentioned in Xilinx public documents, unlike the AES-256 key, there is no storage place for the HMAC key on the FPGA. The HMAC key is instead included in the bitstream. Our investigations revealed that the first encrypted blocks (of the encrypted bitstream) carry the HMAC key. However, since the authentication (examining the correctness of the HMAC) is performed when all bitstream blocks are transferred and decrypted (i.e., at the end of the configuration process), it does not harm the *chosen ciphertext* measurement scenario explained above. Further, after recovering the AES-256 key, the first two blocks of an original bitstream can be decrypted to derive the 256-bit HMAC key. In short, the integrated authentication scheme of all 7 series devices does not have any effect on the efficiency of our presented attack.

4   Conclusions

This work extended the known SCA attacks on the bitstream encryption feature of Xilinx. By means of a sophisticated measurement scenario, i.e., chosen ciphertext, we could reduce the search space from 2^{32} to 2^8 for each step of the attack. This allows the attacks to be mounted by ordinary processing units, e.g., workstation PCs. We have also shown that in case of real-world attacks, the EM analysis using common ordinary EM probes are also possible, where – in contrast to all previous attacks on similar devices based on power consumption – modification of the PCB (where the FPGA is embedded) is not required. Although we have not presented the result of the attacks on Virtex-4 devices, all

FPGA families from 4, 5, 6, and 7 (where the same AES-256 decryption module is integrated) are vulnerable to the attacks presented here.

We should refer to the design and architecture of more recent Xilinx families UltraSCALE and UltraSCALE$^+$, where several security features, e.g., DPA countermeasures, have been integrated. Therefore, the attacks presented in this work are not expected to be portable to the new series devices. However, to the best of our knowledge, all 7 series devices (which are still in production) follow the same architecture and design with respect to bitstream encryption, that predicates on their susceptibility to our attacks.

Acknowledgment. The authors would like to acknowledge Alexander Jakimowic and Oliver Mischke for their help with development of the setup. The research in this work was supported in part by the DFG Research Training Group GRK 1817/1.

References

1. Balasch, J., Gierlichs, B., Verdult, R., Batina, L., Verbauwhede, I.: Power analysis of Atmel CryptoMemory – recovering keys from secure EEPROMs. In: Dunkelman, O. (ed.) CT-RSA 2012. LNCS, vol. 7178, pp. 19–34. Springer, Heidelberg (2012)
2. Brier, E., Clavier, C., Olivier, F.: Correlation power analysis with a leakage model. In: Joye, M., Quisquater, J.-J. (eds.) CHES 2004. LNCS, vol. 3156, pp. 16–29. Springer, Heidelberg (2004)
3. Doget, J., Prouff, E., Rivain, M., Standaert, F.: Univariate side channel attacks and leakage modeling. J. Crypt. Eng. 1(2), 123–144 (2011)
4. Durvaux, F., Standaert, F.: From Improved Leakage Detection to the Detection of Points of Interests in Leakage Traces. IACR Cryptology ePrint Archive, Report/536 (2015)
5. Eisenbarth, T., Kasper, T., Moradi, A., Paar, C., Salmasizadeh, M., Shalmani, M.T.M.: On the power of power analysis in the real world: a complete break of the KEELOQ code hopping scheme. In: Wagner, D. (ed.) CRYPTO 2008. LNCS, vol. 5157, pp. 203–220. Springer, Heidelberg (2008)
6. Heyszl, J., Mangard, S., Heinz, B., Stumpf, F., Sigl, G.: Localized electromagnetic analysis of cryptographic implementations. In: Dunkelman, O. (ed.) CT-RSA 2012. LNCS, vol. 7178, pp. 231–244. Springer, Heidelberg (2012)
7. Kocher, P.C., Jaffe, J., Jun, B.: Differential power analysis. In: Wiener, M. (ed.) CRYPTO 1999. LNCS, vol. 1666, p. 388. Springer, Heidelberg (1999)
8. Lemke-Rust, K.: Models and algorithms for physical cryptanalysis. Ph.D. thesis, Ruhr University Bochum, January 2007
9. Liu, J., Yu, Y., Standaert, F.-X., Guo, Z., Gu, D., Sun, W., Ge, Y., Xie, X.: Small tweaks do not help: differential power analysis of MILENAGE implementations in 3G/4G USIM cards. In: Pernul, G., Y A Ryan, P., Weippl, E. (eds.) ESORICS. LNCS, vol. 9326, pp. 468–480. Springer, Heidelberg (2015). doi:10.1007/978-3-319-24174-6_24
10. Mangard, S., Oswald, E., Popp, T.: Power Analysis Attacks - Revealing the Secrets of Smart Cards. Springer, New York (2007)
11. Moradi, A., Barenghi, A., Kasper, T., Paar, C.: On the vulnerability of FPGA bitstream encryption against power analysis attacks: extracting keys from Xilinx Virtex-II FPGAs. In: Computer and Communications Security, CCS, pp. 111–124. ACM (2011)

12. Moradi, A., Kasper, M., Paar, C.: Black-box side-channel attacks highlight the importance of countermeasures. In: Dunkelman, O. (ed.) CT-RSA 2012. LNCS, vol. 7178, pp. 1–18. Springer, Heidelberg (2012)
13. Moradi, A., Oswald, D., Paar, C., Swierczynski, P.: Side-channel attacks on the bitstream encryption mechanism of AlteraStratix II: facilitating black-box analysis using software reverse-engineering. In: FPGA, pp. 91–100. ACM (2013)
14. Oswald, D., Paar, C.: Breaking Mifare DESFire MF3ICD40: power analysis and templates in the real world. In: Preneel, B., Takagi, T. (eds.) CHES 2011. LNCS, vol. 6917, pp. 207–222. Springer, Heidelberg (2011)
15. Rao, J.R., Rohatgi, P., Scherzer, H., Tinguely, S., Attacks, P.: Or how to rapidly clone some GSM cards. In: IEEE Symposium on Security and Privacy, pp. 31–41. IEEE Computer Society (2002)
16. Schindler, W., Lemke, K., Paar, C.: A stochastic model for differential side channel cryptanalysis. In: Rao, J.R., Sunar, B. (eds.) CHES 2005. LNCS, vol. 3659, pp. 30–46. Springer, Heidelberg (2005)
17. Swierczynski, P., Moradi, A., Oswald, D., Paar, C.: Physical security evaluation of the bitstream encryption mechanism of Altera Stratix II and Stratix III FPGAs. TRETS **7**(4), 34:1–34:23 (2015)
18. Xilinx (Kyle Wilkinson): 7 Series FPGAs Configuration User Guide (2015). http://www.xilinx.com/support/documentation/user_guides/ug470_7Series_Config.pdf
19. Xilinx (Kyle Wilkinson): Using Encryption to Secure a 7 Series FPGA Bitstream (2015). http://www.xilinx.com/support/documentation/application_notes/xapp1239-fpga-bitstream-encryption.pdf
20. Zhou, Y., Yu, Y., Standaert, F.-X., Quisquater, J.-J.: On the need of physical security for small embedded devices: a case study with COMP128-1 implementations in SIM cards. In: Sadeghi, A.-R. (ed.) FC 2013. LNCS, vol. 7859, pp. 230–238. Springer, Heidelberg (2013)

Dismantling Real-World ECC with Horizontal and Vertical Template Attacks

Margaux Dugardin[1,2]($\boxtimes$), Louiza Papachristodoulou[3], Zakaria Najm[1], Lejla Batina[3], Jean-Luc Danger[1], and Sylvain Guilley[1]

[1] COMELEC, TELECOM ParisTech, 46 rue Barrault, 75014 Paris, France
{margaux.dugardin,zakaria.najm,jean-luc.danger,
sylvain.guilley}@telecom-paristech.fr
[2] Thales Communications & Security, CESTI,
3 avenue de l'Europe, 31000 Toulouse, France
[3] Digital Security Group, Radboud University Nijmegen,
P.O. Box 9010, 6500 GL Nijmegen, The Netherlands
louiza@cryptologio.org, lejla@cs.ru.nl

Abstract. Recent side-channel attacks on elliptic curve algorithms have shown that the security of these cryptosystems is a matter of serious concern. The development of techniques in the area of Template Attacks makes it feasible to extract a 256-bit secret key with only 257 traces. This paper enhances the applicability of this attack by exploiting both the horizontal leakage of the carry propagation during the finite field multiplication, and the vertical leakage of the input data. As a further contribution, our method provides detection and auto-correction of possible errors that may occur during the key recovery. These enhancements come at the cost of extra traces, while still providing a practical attack. Finally, we show that the elliptic curve algorithms developed for PolarSSL, and consequently mbedTLS, running on an ARM STM32F4 platform is completely vulnerable, when used without any modifications or countermeasures.

Keywords: Side-channel analysis · Horizontal leakage · Vertical leakage · Scalar multiplication · Brainpool curves · NIST curves · mbedTLS

1 Introduction

Implementing security protocols for embedded devices is a constant challenge for the cryptographic community, due to the development of new and powerful

---

This work was supported in part by the Technology Foundation (STW) through project 12624-SIDES, 13499-TyPhoon (VIDI project) the ICT COST action IC1204 TRUDEVICE and the COST action IC1306 Cryptography for Secure Digital Interaction, Date: 2016-03-04.

F.-X. Standaert and E. Oswald (Eds.): COSADE 2016, LNCS 9689, pp. 88–108, 2016.
DOI: 10.1007/978-3-319-43283-0_6

side-channel attack techniques. By measuring the power consumption or the electromagnetic emanation of a device during the execution of a cryptographic algorithm, it is possible to derive secret data from a single or multiple traces.

Within the area of side-channel attacks there exist different methods of analysis; either by using a single trace (Simple Analysis) or a large number of traces (Differential and Correlation Analysis) from the target device [7,23]. Template Attacks belong to yet another kind of attacks and are considered to be the most powerful method from the information-theoretic point of view, since they take advantage of most information available in a side-channel observation [9]. The attacker is assumed to have one or limited number of side-channel measurements from the target device, i.e. power, EM traces or timing, but he has access to a similar device, on which he can simulate the computations of the target (template building phase). Rechberger and Oswald presented the first practical template attack on RC4 running on an 8-bit micro-controller in [29]. Most notably, the work of De Mulder et al. [27] showed the first practical attack using electromagnetic emanation of Elliptic Curve Cryptosystems (ECC) on an FPGA implementation.

The main idea of the attack presented in this paper is a collision attack exploiting the doubling operation during an ECC computation. Collision attacks exploit the leakage between two portions of the same or different traces, when the same intermediate values are reused. In [4,10,11,15], these attacks are presented as theoretical *horizontal* attacks using collisions. Our work is a practical horizontal attack. The idea of attacking the doubling operation in the elliptic curve setting was originally proposed by Fouque and Valette in [14]. Their "Doubling Attack" is based on the fact that similar intermediate values may be manipulated when working with points $\mathcal{P}$ and $2\mathcal{P}$. However, in most cases, the intermediate values during ECC scalar multiplications are different than the input point. The most efficient result in practical template attacks on ECC is the Online Template Attack (OTA), presented in [3], which requires one full target trace and one template trace per key-bit. With 256 templates, Batina et al. retrieve a 256-bit key on the twisted Edwards curve used for the Ed25519 signature scheme [16].

Our Contribution. In this work, we present a generic attack to scalar multiplication algorithms. Our attack affects the open-source libraries mbedTLS (formerly PolarSSL) and OpenSSL, designed initially for servers and PCs, but easy to adapt to embedded environments, like smart-phones. We demonstrate an attack on PolarSSL v1.3.7 with Brainpool brainpoolP256r1 and NIST P-256 curves running on an ARM Cortex-M4 micro-controller on a STM32F4 platform [25]. The vulnerabilities in the modules of the implementation that make our attack possible, are not fixed in newer versions of PolarSSL (recently bought by ARM and renamed to *mbedTLS* [24]). Therefore, our attack can be applied in the same way as demonstrated in this paper to the most recent version of mbedTLS v.2-2-0.

For the demonstration of our attack, we extend the idea of Online Template Attacks (OTA) by Batina et al. presented in [3]; the authors used one full target trace and one template trace per key-bit to determine the scalar bit-by-bit. However, this method requires an identification phase, in order to compute the threshold between matching and non-matching templates. In our case, there are two different leakage models, a horizontal and a vertical one. Therefore, a threshold as described in [3] cannot be established. We propose a more generic method to distinguish the matching templates.

The horizontal leakage in mbedTLS is a consequence of the software implementation during multiplication of large numbers (256-bit field elements). In most cases, the multiplication of large numbers leaks due to the potential propagation of carry. This carry occurs during the register addition between two words (defined by the length of register). We observed that in OpenSSL (a widely used open-source library) the different propagation of carry leaks in the same way as in mbedTLS, making this library vulnerable not only to our attack, but also to more trivial timing attacks. The timing side-channel leakage due to different propagation of carry can be eliminated by using a dummy operation, such as addition by zero. However, in side-analysis an addition by zero can be detected using vertical leakage. Therefore, this method may create a constant time implementation, but it is still not really efficient to avoid the problem of the propagation of carry. Performing the multiplication in the right-to-left (or from the least significant word to the most significant word) is considered to be a more natural way of multiplication and at the same time more resistant against side-channel attacks.

The vertical leakage that we exploited comes from the Hamming weight of the value stored in the register or the Hamming distance between two values stored in the same register. Despite the fact that the levels of noise (from the USB power supply and the general purpose input/output slots) on the STM32F4 platform are high, the vertical leakage can still be exploited to retrieve the scalar bits.

In the original OTA, the authors did not consider the success rate of the attack and the fact that OTA can fail in recovering the scalar, if one key-bit guess is wrong. If a wrong key-bit assumption cannot be detected, then the error will propagate and the scalar cannot be recovered. Therefore, the advantage of our adaptive template attack over the original OTA is the fact that it detects and corrects errors. Making one assumption for each key-bit and deciding according to the established threshold if this bit is the correct one, does not always give the correct result. In some instances of our attack, the templates obtained for a "0" key-bit assumption was very similar to the template made for the assumption that the key-bit is "1". To increase the success rate of our attack and to determine wrong assumptions, we decided to obtain two template traces for each key-bit. The choice to use both assumptions to create template traces allows to detect and correct any possible error to get back the whole scalar. An efficient way to thwart our attack is to use the countermeasure of randomizing the input point. This is implemented in mbedTLS, but disabled or deterministic by default in the software, because it requires the use of a random generator function from the devices under attack.

Organization of the Paper. This paper is organized as follows: We describe the elliptic curves and the scalar multiplication algorithms used for our attack in Sect. 2. Section 3 gives an overview of the OTA methodology with vertical and horizontal leakage. Section 4 presents our practical adaptive template attack on Brainpool and NIST curves on a STM32F4 micro-controller and the error correction technique used to improve OTA. Section 5 proposes a discussion about the efficiency of certain countermeasures against our attack. Finally, Sect. 6 summarizes our results and concludes the paper.

2 Mathematic Background

2.1 Preliminaries on Elliptic Curves

In 1985, Miller [26] and Koblitz [22] introduced the use of elliptic curves for asymmetric cryptography. The main advantage of using Elliptic Curve Cryptography (ECC) over RSA is the memory and the length of the computations; two important factors for embedded devices. The curves defined over a 256-bit field provide security level of 128-bits, which is equivalent to a 3072-bit RSA key [6].

An elliptic curve $\mathcal{E}$ over $\mathbb{F}_p$ can be defined in terms of solutions to the reduced Weierstrass equation $y^2 = x^3 + ax + b$ over $\mathbb{F}_p$. The pairs (x, y) that verify the previous equation represent the affine coordinate of a point over the curve $\mathcal{E}$. For our experiments, we used the Brainpool curve brainpoolP256r1 recommended by BSI [8] (noted brainpoolP256r1) and the NIST curve P-256 recommended by the NIST standard [28]. These curves are defined over a 256-bit field and have security level of 128 bits (see [17] for more details).

2.2 Scalar Multiplication Algorithm

The scalar multiplication is the main operation in cryptographic protocols using ECC, such as ECDSA signatures [1] or the Diffie-Hellman key-exchange protocol (ECDH) [2]. This is an expensive operation that a designer wants to optimize, yet at the same time a sensitive operation, because it manipulates the secret key. Many scalar multiplication algorithms are used for efficiency and/or resistance against side-channel attacks. In this paper, we perform an attack against the binary left-to-right double-and-add-always algorithm (see in [13,19]), which is considered to be resistant against simple power analysis (SPA). Our attack applies to other regular algorithms as well, similarly to the original OTA [3].

The double-and-add-always algorithm takes as input a point $\mathcal{P} = (x_\mathcal{P}, y_\mathcal{P})$ in affine coordinates and the scalar k. For our experiments the scalar is 256-bit long. For every iteration the computation block performs a doubling operation and an addition with $\mathcal{P}$.

2.3 Scalar Multiplication Module of mbedTLS

MbedTLS is an open-source cryptographic library [24] recently acquired by ARM. MbedTLS contains C and assembly code to speed up the computations over the finite field. The source code is nicely decomposed into modular blocks and it can be used in embedded devices. For ECC operations, it uses the module ecp. To be more efficient the main functions used are doubling in Jacobian coordinates (DBL) and mixed addition [12] between a point in Jacobian and a point in affine coordinates (ADD). The cost of these operations is explained and detailed by Bernstein, Lange et al. in [5]. MbedTLS is intended to be used in embedded systems which include a hardware multiplier, like smart-phones. Scalar multiplication in mbedTLS consists of two steps: multiplication, then modular reduction.

3   Attack Description

3.1 The Main Idea of Online Template Attacks

Online Template Attacks (OTA) constitute an adaptive template attack technique. The main difference with the classical template attacks as described in [9,29] is the absence of the building phase. The attack on the *double-and-always-add left-to-right* algorithm consists of:

1. The attacker first obtains a target trace with input point $\mathcal{P}$ from the target device.
2. The most significant bit is: $k_{MSB} = 1$.
3. To find the bit k_{MSB-i} knowing the previous bits $m_i = k_{MSB} \ldots k_{MSB-i-1}$:
   (a) Template building phase: he obtains two template traces with the input points $[2m_i]\mathcal{P}$ and $[2m_i + 1]\mathcal{P}$.
   (b) Template matching phase: he compares the correlations between the target and each pair of template traces.
   (c) Key bit estimation: the correct key-bit guess is most likely to be given by highest correlation.

The adversary, depending on the implementation and algorithm used, tries to find interesting parts of the trace, where information can leak. In ECC implementations, the interesting parts are inside the scalar multiplication routine and more precisely the doubling and adding operations. For more details on OTA, see the Appendix A or [3] for another scalar multiplication algorithms. The main difference between our attack and the attack described by Batina et al. [3] is that we use two template traces $[2m_i]\mathcal{P}$ and $[2m_i + 1]\mathcal{P}$ to retrieve one key-bit e.g. $2\mathcal{P}$ and $3\mathcal{P}$ for the first bit. By using more template traces, we can detect and correct an error to increase the success rate of the attack.

3.2 Horizontal Leakage Due to Propagation of Carry

Horizontal leakage usually occurs when there are conditional statements in the algorithm. This is the case for PolarSSL v1.3.7, mbedTLS v2.2.0 and OpenSSL v1.0.2 and earlier versions. For mounting our attack, we focus in the doubling operation inside the scalar multiplication. This is the interesting operation that we trigger and create our templates from. In case the whole doubling operation is used to construct templates, it is not possible to achieve high similarity between our templates and the target, mainly due to the noise and the non-constant time implementation. As explained later in Sect. 4.3, we cannot use the intermediate values (in Jacobian coordinates) as input point (in affine) for the templates. However, by focusing on the operations in the first doubling of the double-and-add-always algorithm to construct the template traces, we achieve more accurate results. For the template pattern, we need only the pattern of the first finite-field multiplication in the doubling. In our implementation (Algorithm 13 in [30]), the first operations during the doubling of point $\mathcal{P} = (X, Y, Z)$ are the following[1]:

$$D_1 \leftarrow X \times X \qquad \mathrm{mod}\ p$$
$$D_2 \leftarrow Y \times Y \qquad \mathrm{mod}\ p \tag{1}$$
$$\vdots$$

In PolarSSL v1.3.7 and mbedTLS v2.2.0 a multiplication between two elements in the finite field is computed as described in Algorithm 1. The result of the multiplication is stored in a 512-bit element, called *"multiplication-before-reduction"*; then the result is reduced modulo p (the characteristic of the finite field). For the curves defined in Sect. 2, one element in the finite field is 256-bit long. The micro-controller is Cortex-M4 with 32-bit registers for data operations (see Sect. 4.1 for more details). Therefore, one field element corresponds to 8 words of 32-bits.

Let A and B be two 256-bit elements in the finite field. Then, A (resp. B) can be written as 8 words A_i for all $i \in \{0, 1, \ldots, 7\}$ (resp. B_i) of 32-bits. A_0 is the least significant word (LSW) of A and A_7 is the most significant word (MSW) of A. Let X be the result of the multiplication $A \times B$ before reduction; X can be represented by 16 words of 32-bits $(X_{15}X_{14}\ldots X_0)$. The Algorithm 1 shows how multiplication is performed in mbedTLS. The result $A \times B_i$ is stored in eight 32-bit words and there is a potential carry C, which needs to be stored separately (see step 3). This potential overflow creates a significant pattern that can be distinguished from its high amplitude when $C = 1$; this pattern is the propagation of carry as depicted in Fig. 1. The leakage due to the propagation of carry depends on the MSW of the input data A_7. For brainpoolP256r1, $\max\{A_7 | A \in \mathbb{F}_p\} = \texttt{0xA9FB57DA}$ and the probability of

---

[1] The beginning of the doubling operation is the implementation in PolarSSL v1.3.7. The sequence of the finite field operations in the doubling operation in the mbedTLS v2.2.0 changes to: $D_1 \leftarrow X \times X, D_2 \leftarrow 3 \times X$, but this does not affect the efficiency of our attack.

Algorithm 1. Multiplication in mbedTLS

Require: A and $B_7..B_0$ two elements of 256-bits long.
Ensure: $X = A \times B$
1: $X \leftarrow 0$
2: **for** i from 7 down to 0 **do**
3:    $(C, X_{i+7}, X_{i+6}, \ldots, X_i) \leftarrow (X_{i+7}, \ldots, X_i) + A \times B_i$
4:    $j \leftarrow i + 8$
5:    **repeat**
6:       $(C, X_j) \leftarrow X_j + C$
7:       $j \leftarrow j + 1$
8:    **until** $C \neq 0$
9: **end for**
10: **return** X

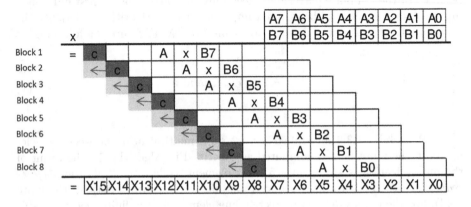

Fig. 1. Propagation of carry during multiplication in the field

having a propagation of carry is close to $p = 0.17$. For P-256, $\max\{A_7 | A \in \mathbb{F}_p\}$ equals $2^{32} - 1$, so this probability is close to $p = 0.25$. The full proof of this computation of probability is described in Appendix B. As shown in Fig. 1, we can have 7 propagations during the multiplication, but we cannot detect the last propagation. So, the probability to have two templates with the same propagation of carry, denoted by $\mathbb{P}(C)$, is:

$$\mathbb{P}(C) = \sum_{i=0}^{6} \binom{6}{i} p^{2i}(1 - p)^{2(6-i)} \tag{2}$$

where p is the probability to have an internal propagation of carry. The probability to have horizontal leakage is 0.86 using $p = 0.17$ for brainpoolP256r1. For P-256, the probability to have horizontal leakage is 0.95 using $p = 0.25$.

However, it is more interesting from the OTA point of view to find out when a difference in the propagation of carry occurs between the target and template traces. This is the only part of mbedTLS that is non-constant time and we take advantage of this timing difference, every time it occurs. In this case, there is

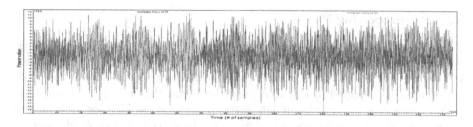

Fig. 2. Squaring of two random data with different propagation of carry

an obvious horizontal leakage between the target and the template traces, as depicted in Fig. 2.

3.3 Vertical Leakage Due to Signal Amplitude

In constant time executions of our implementation, there is no difference in the propagation of carry and the template traces are synchronized with the target trace. In those cases, we observe only a vertical leakage due to the amplitude of the signal and the same method as described in [3] can be used. To observe this leakage, we use the pattern matching technique using the Pearson correlation as a distinguisher.

4 Detailed Phases of the Attack in Practice

4.1 Acquisition Setup

The target device is an STM32F4 micro-controller, which contains an ARM Cortex-M4 processor running at its maximum frequency (168 MHz). We imported the assembly code originally included in PolarSSL v1.3.7 to ARM Cortex-M4 and implemented the double-and-add-always procedure as described in [13,19]. For the acquisition, we used a 54855 Infiniium Agilent oscilloscope and a Langer EMV-TECHNIK RF-U5-2 near field probe. The sampling frequency is 1 GSa/s with 50 MHz hardware input low-pass filter enabled. Matlab 2014b is used for the analysis, and Inspector SCA tool [18] for depicting the traces in this paper. The position of the probe was determined to maximize the signal related to the activity of the 32×32 hardware multiplier[2].

For the curves defined in Sect. 2, one element in the finite field is 256-bit long. Thus, each operation over the field consists of manipulating eight processor words (8×32 bits). In our implementation, a multiplication-before-reduction consists of eight multiplications between a 256-bit element by each 32-bit words of the second element. It leads to eight easily identifiable patterns of eight blocks on EM traces. The length between two blocks can be different depending on the propagation of carry, as explained in Sect. 3.2.

---

[2] This is a simple identification phase, where we scan the device and find where the crypto processor is. Then we just move the probe around this position, in order to get a signal as clear as possible.

4.2 Pre-processing Phase

The pre-processing phase starts with choosing an input point $\mathcal{P}$ and obtaining the target trace from our target device; this is depicted in Fig. 3. In this trace, we need to spot the multiplication patterns, which are eight blocks of 256×32 multiplications depicted in Fig. 4. We note here that this does not constitute a building phase in the usual template setting, it is just an identification phase. From the implementation and the device, we know that a 256-bit element is processed into 32-bit multipliers. Therefore, we expect to see eight patterns for each multiplication. The multiplication procedure is described in Sect. 3.2.

When we have a clear pattern for the multiplication, we cross correlate this pattern with our target trace and we obtain the cross-correlation pattern with one peak at the position of every multiplication. Figure 5 shows the cross correlation of the target trace with the multiplication pattern. By counting the peaks in the cross-correlation trace, we can find the part of the computation that we are interested in. For brainpoolP256r1, as explained in [5], the doubling consists of 10 multiplications (except for the first doubling, where there are only 7 multiplications[3]), and the mixed addition consists of 11 multiplications. For P-256, there is a particular parameter equal to $(-3 \mod p)$, so the multiplication by a in the doubling can be optimized. The doubling consists of 9 multiplications and the mixed addition of 11 multiplications[4].

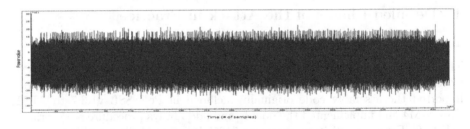

Fig. 3. EM acquisition for scalar multiplication on P-256 with $k = \texttt{0xA5A5}$

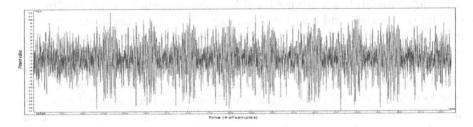

Fig. 4. Pattern of multiplication-before-reduction

---

[3] Because in the beginning $Z = 1$ and we computed aZ^4 with 3 multiplications.

[4] The fact that doubling is performed faster for P-256, allows us to recover 7 bits of the scalar at once.

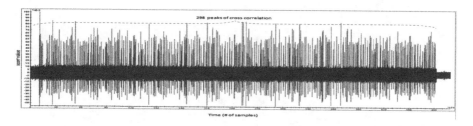

Fig. 5. Cross correlation between the pattern of the multiplication and the target trace

In this way, we can "cut" the target trace in sections according to the loop of the scalar multiplication operation (as in Fig. 6). The first iteration of the double-and-add always algorithm is completed after 18 peaks of cross-correlation. For the next iterations, we take into account that each doubling consists of 9 or 10 multiplications and each addition of 11. For the first bit, the interesting section on the target is the 19th multiplication. For the second bit, the interesting section is the 39th multiplication for P-256 or the 40th multiplication for brainpoolP256r1. For the third bit, the interesting section on the target is the 59th multiplication for P-256 or the 61th multiplication for brainpoolP256r1, and so on for all the other bits of the scalar.

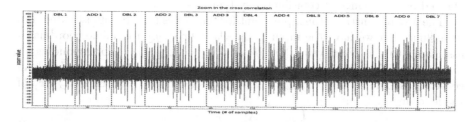

Fig. 6. The first seven iterations of the scalar multiplication algorithm on the curve

As the last step of this phase, we calculate multiples of the point $\mathcal{P}$ using our PolarSSL v1.3.7 implementation. We explain this in detail in the next section.

4.3 Template Acquisition

In PolarSSL v1.3.7 (and generally mbedTLS) every input point is represented in affine coordinates and then converted to Jacobian coordinates. The intermediate values are represented in Jacobian coordinates. The input points to the device are given in affine coordinates. To create templates, we need to find an input point in affine corresponding to an intermediate value in Jacobian coordinates.

The *target trace* is obtained with input point $\mathcal{P} = (x_{\mathcal{P}}, y_{\mathcal{P}})$ given in affine coordinates. In order to compute the intermediate values of the points $[2m_i]\mathcal{P} =$

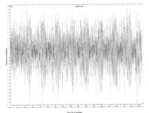

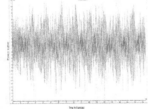

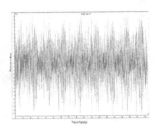

Fig. 7. Pattern of the 19th multiplication in trace with input $\mathcal{P}$

Fig. 8. Pattern of the 1st multiplication in trace with input $\mathcal{Q}_0$

Fig. 9. Pattern of the 1st multiplication in trace with input $\mathcal{Q}_1$

$(X_{[2m_i]\mathcal{P}}, Y_{[2m_i]\mathcal{P}}, Z_{[2m_i]\mathcal{P}})$ and $[2m_i + 1]\mathcal{P} = (X_{[2m_i+1]\mathcal{P}}, Y_{[2m_i+1]\mathcal{P}}, Z_{[2m_i+1]\mathcal{P}})$ with PolarSSL v1.3.7, we use the formulas defined in [5]. Note that this does not correspond to the point $[2m_i]\mathcal{P}$ and $[2m_i + 1]\mathcal{P}$ in affine coordinates, because $Z_{[2m_i]\mathcal{P}}, Z_{[2m_i+1]\mathcal{P}} \neq 1$. Therefore, we cannot compare directly the templates with input point $[2m_i]\mathcal{P}$ (resp. $[2m_i + 1]\mathcal{P}$), since they are not in affine form.

We create our templates with a specific input point $\mathcal{Q}_i$ such that the first field multiplication D_1 in $2\mathcal{P}$ or $3\mathcal{P}$ is the same with the one attacked on the target trace. The squaring of the X-coordinate of the intermediate value is not affected by the change of coordinates system.

The way to construct the input point for templates is more sophisticated. Let us assume that we have the input point $\mathcal{Q}_0 = (x_{\mathcal{Q}_0}, y_{\mathcal{Q}_0})$ in affine coordinates associated to the point value $[2m_i]\mathcal{P}$ and $\mathcal{Q}_1 = (x_{\mathcal{Q}_1}, y_{\mathcal{Q}_1})$ corresponding to $[2m_i + 1]\mathcal{P}$. We need to analyse the squaring of the X-coordinate in Jacobian coordinates. The input point $\mathcal{Q}_0 = (x_{\mathcal{Q}_0}, y_{\mathcal{Q}_0})$ should be a solution in $\mathbb{F}_p \times \mathbb{F}_p$ of the following system:

$$\begin{cases} x_{\mathcal{Q}_0} = X_{[2m_i]\mathcal{P}} \\ y_{\mathcal{Q}_0}^2 = x_{\mathcal{Q}_0}^3 + a x_{\mathcal{Q}_0} + b \end{cases} \tag{3}$$

with a, b the parameters of the curve as defined in [8,28].

The number X used as input in the squaring is random, so $X^3 + aX + b$ is also random. If $x_{\mathcal{Q}_0}^3 + a x_{\mathcal{Q}_0} + b$ is not a square in the finite field, we can change one bit in X as proposed in [3] we get another point on the curve that satisfies Eq. (3).

We locate the first multiplication in the template trace corresponding to the squaring of the X-coordinate of the input point $\mathcal{Q}_0$ or $\mathcal{Q}_1$, depicted in Figs. 8 and 9 respectively. With these two patterns and the target trace (Fig. 7), we can perform template matching.

4.4 Template Matching

In this section, we present how to perform template matching by making the right hypothesis on a scalar-bit. This procedure is described for the cases of horizontal and vertical leakage. The probability of having a horizontal leakage corresponds to the probability of having different propagation of carry between the two templates.

Horizontal Leakage. When the traces are not synchronized (86 % of cases in brainpoolP256r1 curve, and 95 % in P-256), cross correlation between the multiplication pattern and the target trace is performed before template matching, in order to choose the correct part of the target trace. Then we align the template and target traces and decide what the correct key-bit guess is.

Horizontal leakage is observed when there is different propagation of carry between two multiplications 256×32 in the field. In Fig. 10 we see the misalignment of the traces due to propagation of carry.

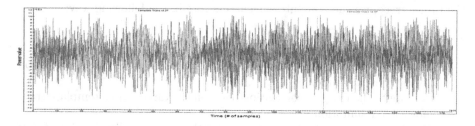

Fig. 10. Misalignment of two template traces due to propagation of carry

Vertical Leakage. When the implementation is executed in constant time and the template traces are synchronized with the target trace, the same method as described in [3] can be used (14 % of cases in brainpoolP256r1 and 5 % in P-256). The propagation of carry is the same between the two templates and the target as depicted in Fig. 11; therefore, we can only observe a *vertical* leakage in our traces. In our experiments, we use the Pearson correlation coefficient $\rho(X, Y)$ as described in Appendix A.3 and we get a correlation of $0, 81$ for the multiplication obtained from the target trace and the template trace of $\mathcal{Q}_0$. The same value drops to $0, 78$ for the correlation of the target trace with the template trace of $\mathcal{Q}_1$.

4.5 Success Rate for One Key-Bit

In this part, the method used to calculate and increase the success rate of our attack is described. As explained in Sect. 3.2, the probability to have a different propagation of carry between two templates is 95 % for P-256 and 86 % for brainpoolP256r1. The horizontal attack scenario is easy, since if two templates have different propagation of carry, then the success rate of finding this bit is 100 %.

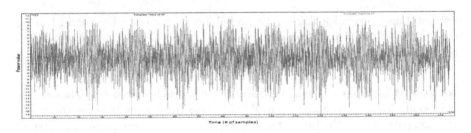

Fig. 11. Two templates with the same propagation of carry

For the vertical attack scenario, the success rate depends on the input data. Therefore, we examine only the input data, for which the propagation of carry is the same in two template traces. By using random points on each curve for the target trace, we can compute the success rate in the following way:

1. Acquire 30 target traces.
2. For each target traces and different key bit value:
   (a) Acquire 100 template traces for each assumption when the propagation of carry is the same.
   (b) Compute the Pearson correlation between the target and a template trace for each assumption.
   (c) With k_i we denote the correct guess for the corresponding key-bit of k. If k_i gives the highest Pearson correlation, then the counter corresponding to the success of the attack increases. If $\neg k_i$ has higher correlation, then the counter corresponding to the failure of the attack increases.

The success rate per key-bit in vertical leakage is 76.23 % for P-256 and 69 % for brainpoolP256r1. The total success rate to find one key-bit, independent of the leakage model, is $1 \times 0.95 + 0.76 \times 0.05 = 98.8\%$ for P-256, and $1 \times 0.86 + 0.69 \times 0.14 = 95.66\%$ for brainpoolP256r1.

Averaging template traces can increase the vertical information leakage. When the scalar is randomized, we cannot perform the attack with more than one target traces. However, we can still acquire more than one template traces. By using only one target trace with an average of a few template traces, the success rate increases as shown the Table 1 on the brainpoolP256r1 curve. For instance, by using 100 template traces the success rate for brainpoolP256r1 curve is $1 \times 0.86 + 0.99 \times 0.14 = 99.86\%$.

Table 1. Different success rates according to the number of average template traces on brainpoolP256r1 curve.

Number of average traces	1	10	50	100
Success rate		69 %	80.70 %	91.60 %	99.80 %

4.6 Error-Correcting Bit from the Template Traces

The novelty of our method is the possibility of detection and correction of errors. As we described in the previous section, the success rate to retrieve one bit using OTA is close to 99 %, which means that there is a 1 % probability of having an unsuccessful attack due to a wrong key-guess. For a 256-bit scalar, if an error occurs in the beginning and it is not detected, the success rate for the original OTA is 7.6 % ($0.99^{255} \simeq 0.076$), since this error will propagate and affect all the bits after the wrong guess. Therefore, it is very important to detect and correct errors before making new templates. An error can be made when both template traces have the same propagation of carry. In order to be sure for a key-bit value, the value of this current bit and the following one can be computed. For instance, if the two templates for $2\mathcal{P}$ and $3\mathcal{P}$ have the same propagation of carry, then we create templates for $4\mathcal{P}$, $5\mathcal{P}$, $6\mathcal{P}$ and $7\mathcal{P}$. The following four cases can occur:

1. One template has the same propagation with the target: Then, we can determine correctly both key-bits. For instance, if the template trace of $7\mathcal{P}$ gives the same propagation of carry as the target trace, then the only possible bit values are "1" for the second key-bit and "1" for the third one.
2. Two templates have the same propagation with the target: We need to compute the next 4 templates ($4i, 4i + 1, 4i + 2, 4i + 3$, where k_i is the i-th bit of the exponent), and recover the next bit using these 4 template traces.
3. Three templates have the same propagation with the target: We compute 6 templates, and recover the next bit using those 6 template traces.
4. Four templates have the same propagation with the target: We compute 8 templates, and recover the next bit using those 8 template traces.

The probability to have one template trace with the same propagation of carry with the target trace on brainpoolP256r1 is 70 %, two template traces is 14 %, 3 template traces is 3.9 %, 4 template traces is 1.2 %, 5 template traces is 0.4 %, 6 template traces is 0.1 %. The probability for more template traces with the same propagation of carry is very low. For P-256, the probability to have one template trace with the same propagation of carry with the target trace is, 90 %, two template traces is 5.9 %, 3 template traces is 0.7 %, 4 template traces is 0.1 %, 5 template traces is 0.01 %. For both curves, this probability reduces significantly for more template traces. As a conclusion, the number of template traces cannot increase exponentially. At the end of the attack, in order to retrieve the 256-bit scalar, there can be an uncertainty for the last 2 or 3 bits. By exhaustive key search or by comparing the corresponding templates with the template of $\mathcal{Q} = [k]\mathcal{P}$, the last 2^2 or 2^3 scalar key-bits can be found.

5 Countermeasures

At this point, it is clear that both OTA and our adaptive template attack are very efficient methods to attack scalar multiplication during the execution of ECC protocols. These methods can be easily adapted to other scalar multiplication algorithms as described in [19,30]. For the binary algorithm Montgomery

Ladder [21], we can create templates for the doubling operation and find the correct key-bit. For the non-binary algorithm using windows, we can obtain templates for all hypotheses and make the same attack with more template traces. Since the most commonly used scalar multiplication algorithms are vulnerable to our attack, it is interesting to see which countermeasures can be applied against it. We hereby give of the classical countermeasures against side-channel attacks and their efficiency against our attack:

– *Randomization of the scalar.*
The result of randomizing the scalar will be getting traces with $[k']P$ instead of $[k]P$, with k' defined below. The important property that thwarts randomization of the scalar, is the fact that we need only one target trace, this the same randomized k' is manipulated throughout the attack. For the template traces, we always need the first part of the trace, which corresponds to the beginning of the scalar multiplication algorithm running with input point a multiple m of P. This part of the trace is not affected by the randomization of the scalar. The different ways of scalar randomization are:
1. $[k]\mathcal{P} = [k - r]\mathcal{P} + [r]\mathcal{P}$, two scalar multiplications are computed $\mathcal{Q} = [k - r]\mathcal{P}$ and $\mathcal{R} = [r]\mathcal{P}$;
2. $[k]\mathcal{P} = [k \times r^{-1}]([r]\mathcal{P})$, two scalar multiplications are computed $\mathcal{Q} = [r]\mathcal{P}$ and $\mathcal{R} = [k \times r^{-1}]\mathcal{Q}$;
3. $[k]\mathcal{P} = [k \mod r]\mathcal{P} + [\lfloor \frac{k}{r} \rfloor]([r]\mathcal{P})$, three scalar multiplications are computed $\mathcal{Q} = [k \mod r]\mathcal{P}$, $\mathcal{R} = [r]\mathcal{P}$ and $\mathcal{S} = [\lfloor \frac{k}{r} \rfloor]\mathcal{R}$.
4. $[k]\mathcal{P} = [k']\mathcal{P} - [rq]\mathcal{P}$, where q is the order of the curve and $k' = k + rq$ is the randomized scalar. Thus, $[rq]\mathcal{P} = \mathcal{O}$ the neutral element on $\mathcal{E}$.
For all these randomization techniques, our attack can be applied; the *target trace* requires one acquisition on the second or third scalar multiplication. This acquisition is possible using an oscilloscope with big memory depth. For the template traces, we make assumptions for each part of the scalar multiplication. We retrieve a random scalar part for each scalar multiplication part. In order to retrieve the scalar, we compute the addition (randomization 1.), the multiplication (randomization 2.) or both addition and multiplication of the scalar (randomization 3.). For the fourth case, we can recover the randomized scalar k'. Therefore, scalar randomization is not efficient against our type of attack.
– *Randomization of the point.*
The Jacobian representation of the point can be easily randomized similar to projective coordinates. For Jacobian coordinates, the randomization consists in selecting a random r in the finite field $\mathbb{F}_p$, and computing: $(X, Y, Z) \mapsto (r^2 \times X, r^3 \times Y, r \times Z)$. In most cases, the input point is in affine coordinates, so the randomization of the point is reduced to compute: $(x, y) \mapsto (r^2 \times x, r^3 \times y, r)$. The supplementary cost of this countermeasure is 5 finite field multiplications. For comparison, the cost of one scalar multiplication using 256-bit scalar with a regular algorithm such as double-and-add-always is 5100 multiplications. Applying randomization of the input point does not allow to predict intermediate values of the calculation and prevents

the construction in a deterministic way for the template traces. This counter-measure is implemented in mbedTLS, and it should be used when the device under attack has a random generator.

- *Random isomorphic elliptic curve.*
The idea to protect scalar multiplication by transforming a curve through various random morphisms was initially proposed by Joye and Tymen in [20]. Assume that ϕ is a random isomorphism from $\mathcal{E}_K \rightarrow \mathcal{E}'_K$, which maps $\mathcal{P} \in \mathcal{E}_K \rightarrow \mathcal{P}' \in \mathcal{E}'_K$. Multiplying $\mathcal{P}'$ with k will give $\mathcal{Q}' = [k]\mathcal{P}' \in \mathcal{E}'_K$. With the inverse map ϕ^{-1} we can get back to $\mathcal{Q} = [k]\mathcal{P}$. For applying our attack, we need to know the internal representation of the point, so if $\mathcal{P}'$ is on a different curve that the adversary does not know, he cannot create input points in this representation.

6   Conclusions

In this paper we presented a practical extension of OTA on Brainpool brain-pool256r1 and NIST P-256 curves implemented on an ARM Cortex M4 micro-controller. A modified version of OTA is applied with the Pearson correlation coefficient as distinguisher for the correct hypothesis on the key-bit.

Error detection and correction of a wrong key-bit guess is possible for our adaptive template attack, and increases the success rate of the attack from 7.6 % to 99.8 % for a 256-bit scalar. We achieve these results by averaging 100 template traces and using two template traces to recover each key-bit.

Horizontal leakage due to conditional statements was not expected to be seen in recent cryptographic implementations, but unfortunately they are still used. An implementation without conditional statements would not prevent our attack, but it would reduce its success rate to that of OTA.

Most of the countermeasures applicable to the original OTA attack should also work against our attack. Randomizing the input point, by randomizing its coordinates, for every execution of the attacked algorithm is the most efficient countermeasure against OTA, though incurring with some cost for the perfor-mance of the implementation. Point randomization is also efficient against our attack, since we need to know the input point and its intermediate values. Actu-ally, the adversary needs to be able to choose input points for templates. The countermeasure of point blinding must be activated, in order to use mbedTLS in ARM embedded devices. Adoption of this countermeasure is not straight-forward, because it requires the use of a random generator in the device under attack.

Acknowledgements. The authors would like to thank the anonymous reviewers for their useful comments that improved the quality of the paper. Moreover, the first author would like to thank Jean-Christophe Courrège and Carine Therond for useful comments on an earlier version of this work.

A   Description for Online Template Attack

A.1   Attack Model for OTA

Online Template Attacks (OTA), introduced in [3], is an adaptive template attack technique, which can be used to recover the secret scalar in a scalar multiplication algorithm. The main assumption in the OTA attacker model is in his ability to choose an input point to the scalar multiplication algorithm, in order to generate template traces. As it is demonstrated in the original paper, OTA works with one *target trace* from the device under attack and one *template trace* per key-bit obtained from the attacker's device that runs the same implementation. Performing OTA in practice requires the following assumptions to be made regarding the attacker:

- The attacker knows the input P of the target device.
- He knows the implementation of the scalar multiplication algorithm and he is able to compute the intermediate values.
- He can choose the input points on a device similar to the target device.

Furthermore, we work with the following assumptions related to the device:

- The scalar can be randomized.
- The intermediate values are deterministic.

The OTA is then performed as follows:

1. The attacker first obtains a target trace with input point P from the target device.
2. He obtains template traces with input points $[m]P, m \in \mathbb{Z}$ for multiples of the point P, e.g. $2P$ or $3P$.
3. He compares the correlations between the target and each pair of template traces. The correct guess is most likely to be the highest correlation.

   The OTA technique is originally described for binary algorithms, but it can be easily adapted to the windows method by creating one template for a hypothesis made for each window.

   The attacker model for OTA is more suitable for the Diffie-Hellman key-exchange protocol, because the input point can be selected. Nevertheless, this attack can be applied against the ECDSA algorithm, if the input point of the target device is known.

A.2   Constructing Template Traces for OTA

At this point, it is important to explain precisely how the interesting points to generate the template traces are chosen. With the term *interesting points* we mean the multiples of the point P that are expected to be the outputs of every iteration of the scalar multiplication algorithm, i.e. $2P$ and $3P$ for the first bit of the scalar. This is demonstrated with a graphical example depicted in Fig. 12.

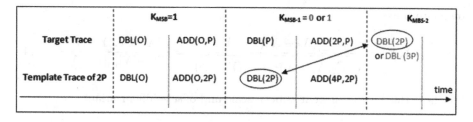

Fig. 12. How to find the second MSB K_{MSB-1} in the target trace with the template trace of $2\mathcal{P}$

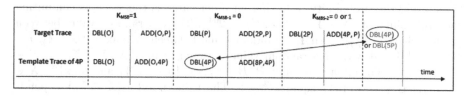

Fig. 13. How to find the third MSB K_{MSB-2} in the target trace with the template trace of $4\mathcal{P}$

Let us assume that the initial input point to the double-and-add-always algorithm is $\mathcal{P}$ and the most significant bit (K_{MSB}) of our secret scalar is 1. Then, the output of the second iteration (operations for K_{MSB-1}) is either $2\mathcal{P}$ or $3\mathcal{P}$. For example, if $K_{MSB-1} = 0$, then the output of the second iteration is $2\mathcal{P}$ and consequently the template trace for $2\mathcal{P}$ gives higher correlation to the target trace than the template for $3\mathcal{P}$. We compute the correlations between the template traces $2\mathcal{P}, 3\mathcal{P}$, and the target trace, in order to find the most likely key-bit. The highest correlation value is considered to be the right key guess.

We continue the same procedure of calculating the two possible outcomes for bit K_{MSB-2}, which are the template traces for $4\mathcal{P}$ or $5\mathcal{P}$, and then finding the highest correlation between the templates and the target trace. Figure 13 shows how the templates for the third bit K_{MSB-2} can be generated. In general, for each iteration of the scalar multiplication algorithm, we compare the second iteration of the scalar multiplication execution (corresponding to the first doubling operation whose consumption is detected with EM) in the template trace with the $(i + 1)^{\text{th}}$ execution of the target trace.

A.3   Template Matching Phase

Template matching is performed at suitable parts of the traces, where key-bit related assignments take place. Our pattern matching technique, in order to distinguish the right hypothesis on the attacked bit of the scalar, is based on the Pearson correlation coefficient $\rho(X, Y)$ between the target trace and the template traces.

$$\rho(X,Y) = \frac{\sum_i (X_i - \bar{X})(Y_i - \bar{Y})}{\sqrt{\sum_i (X_i - \bar{X})^2}\sqrt{\sum_i (Y_i - \bar{Y})^2}} = \frac{\langle X - \bar{X}, Y - \bar{Y}\rangle}{||X - \bar{X}||\ ||Y - \bar{Y}||} \qquad (4)$$

We chose this metric, since it is both scale and offset-shift invariant.

B   Probability of the Propagation of Carry

Computing the probability of having an inner carry is the same as computing the probability of $(X \times Y + R \times 2^{32}) \geq 2^{64}$ with X a random value between $[0, \max\{A_7 | A \in \mathbb{F}_p\}]$, with Y a random value between $[0, \max\{B_i | B \in \mathbb{F}_p, i \in \{0, \cdots, 7\}\}]$ and with R a random value between $[0, \max\{X_i | X \in \mathbb{F}_{(p-1)^2}, i \in \{7, \cdots, 15\}\}]$. For all curves, $\max\{B_i | B \in \mathbb{F}_p, i \in \{0, \cdots, 7\}\}$ and $\max\{X_i | X \in \mathbb{F}_{(p-1)^2}, i \in \{7, \cdots, 14\}\}$ equal $2^{32} - 1$. The value $\max\{A_7 | A \in \mathbb{F}_p\}$ depends on the MSW of the characteristic of the finite field. The probability can be computed as follows:

$$\mathbb{P}(X \times Y + R \times 2^{32}) \geq 2^{64}) = \frac{1}{4}\frac{\max\{A_7 | A \in \mathbb{F}_p\}^2}{2^{64}} \qquad (5)$$

We hereby give a the complete computation of the probability of an inner-carry propagation (Eq. 5)

$$\mathbb{P}(XY + 2^{32}R \geq 2^{64})$$

$$= \sum_{x=0}^{A_7-1}\sum_{y=0}^{2^{32}-1}\sum_{r=0}^{2^{32}-1} \mathbb{P}(XY + 2^{32}R \geq 2^{64} \mid X = x, Y = y, R = r)\mathbb{P}(X = x)\mathbb{P}(Y = y)\mathbb{P}(R = r)$$

$$= \sum_{x=0}^{A_7-1}\sum_{y=0}^{2^{32}-1}\sum_{r=0}^{2^{32}-1} \mathbb{P}(xy + 2^{32}r \geq 2^{64})\frac{1}{A_7}\frac{1}{2^{32}}\frac{1}{2^{32}}$$

$$= \frac{1}{A_7}\frac{1}{(2^{32})^2}\sum_{x=0}^{A_7-1}\sum_{y=0}^{2^{32}-1}\sum_{r=0}^{2^{32}-1} 1_{xy+2^{32}r\geq 2^{64}}, \text{ where 1 is the indicator, i.e., } 1_z = \begin{cases} 0 & \text{if } z \text{ is false,} \\ 1 & \text{otherwise} \end{cases}$$

which can be approximated by:

$$\frac{1}{A_7}\frac{1}{(2^{32})^2}\int_{x=0}^{A_7-1}\int_{y=0}^{2^{32}-1}\int_{r=0}^{2^{32}-1}\delta_{xy+2^{32}r\geq 2^{64}}\mathrm{d}r\mathrm{d}y\mathrm{d}x$$

$$\simeq \frac{1}{A_7}\frac{1}{(2^{32})^2}\int_{x=0}^{A_7}\int_{y=0}^{2^{32}}\int_{r=0}^{2^{32}}\delta_{xy+2^{32}r\geq 2^{64}}\mathrm{d}r\mathrm{d}y\mathrm{d}x$$

$$= \frac{2^{32}}{A_7}\int_{x=0}^{a_7}\int_{y=0}^{1}\int_{r=0}^{1}\delta_{xy+r\geq 1}\mathrm{d}r\mathrm{d}y\mathrm{d}x$$

with $x \leftarrow x/2^{32}$, $y \leftarrow y/2^{32}$, $r \leftarrow r/2^{32}$ and $a_7 = A_7/2^{32}$.

It holds, $\delta_{xy+r\geq 1} = \delta_{r\geq 1-xy}$. Besides, $1 - xy \in [1 - a_7, 1] \subset [0, 1]$. Indeed,

$$0 \leq x \leq a_7, 0 \leq y \leq 1 \implies 0 \leq xy \leq a_7, \text{ hence } 1 - a_7 \leq 1 - xy \leq 1.$$

Therefore,

$$\frac{2^{32}}{A_7} \int_{x=0}^{a_7} \int_{y=0}^{1} \int_{r=0}^{1} \delta_{xy+r\geq 1} \mathrm{drdydx} \quad = \frac{2^{32}}{A_7} \int_{x=0}^{a_7} \int_{y=0}^{1} \int_{r=1-xy}^{1} \mathrm{drdydx}$$

$$= \frac{2^{32}}{A_7} \int_{x=0}^{a_7} \int_{y=0}^{1} xy \mathrm{dydx} \quad = \frac{2^{32}}{A_7} \int_{x=0}^{a_7} x \mathrm{dx} \times \int_{y=0}^{1} y \mathrm{dy}$$

$$= \frac{2^{32}}{A_7} \left[\frac{x^2}{2}\right]_0^{a_7} \times \left[\frac{y^2}{2}\right]_0^{1} \quad = \frac{2^{32}}{A_7} \frac{a_7^2}{2} \times \frac{1}{2} = \frac{2^{32}}{A_7} \frac{1}{4} a_7^2 = \frac{1}{4} \frac{A_7}{2^{32}}.$$

For $A_7 = 2^{32-1}$, this yields ≈ 0.25. For $A_7 = \mathtt{0xA9FB57DA}$, this yields ≈ 0.166.

References

1. ANSI-X9.62. Public Key Cryptography for the Financial Services Industry: The Elliptic Curve Digital Signature Algorithm (ECDSA) (1998)
2. ANSI-X9.63. Public Key Cryptography for The Financial Services Industry: Key Agreement and Key Transport Using Elliptic Curve Cryptography (1998)
3. Batina, L., Chmielewski, L., Papachristodoulou, L., Schwabe, P., Tunstall, M.: Online template attacks. In: Proceedings of Progress in Cryptology - INDOCRYpPT –15th International Conference on Cryptology in India, New Delhi, India, 14–17 December, pp. 21–36 (2014)
4. Bauer, A., Jaulmes, E., Prouff, E., Wild, J.: Horizontal collision correlation attack on elliptic curves. In: Lange, T., Lauter, K., Lisoněk, P. (eds.) SAC 2013. LNCS, vol. 8282, pp. 553–570. Springer, Heidelberg (2014)
5. Bernstein, D.J., Lange, T.: Explicit formulas database. http://www.hyperelliptic.org/EFD/
6. Cryptographic Key Implementation BlueKrypt
7. Brier, E., Clavier, C., Olivier, F.: Correlation power analysis with a leakage model. In: Joye, M., Quisquater, J.J. (eds.) CHES 2004. LNCS, vol. 3156, pp. 16–29. Springer, Heidelberg (2004)
8. BSI: RFC(5639)-Elliptic Curve Cryptography (ECC) Brainpool Standard Curves and Curve Generation. Technical report, Bundesamt für Sicherheit in der Informationstechnik (BSI) (2010)
9. Chari, S., Rao, J.R., Rohatgi, P.: Template attacks. In: 4th International Workshop on Cryptographic Hardware and Embedded Systems - CHES, Redwood Shores, CA, USA, August 13–15, Revised Papers, pp. 13–28 (2002)
10. Clavier, C., Feix, B., Gagnerot, G., Giraud, C., Roussellet, M., Verneuil, V.: ROSETTA for single trace analysis. In: Galbraith, S., Nandi, M. (eds.) INDOCRYPT 2012. LNCS, vol. 7668, pp. 140–155. Springer, Heidelberg (2012)
11. Clavier, C., Feix, B., Gagnerot, G., Roussellet, M., Verneuil, V.: Horizontal correlation analysis on exponentiation. In: Soriano, M., Qing, S., López, J. (eds.) ICICS 2010. LNCS, vol. 6476, pp. 46–61. Springer, Heidelberg (2010)
12. Cohen, H., Miyaji, A., Ono, T.: Efficient elliptic curve exponentiation using mixed coordinates. In: Ohta, K., Pei, D. (eds.) ASIACRYPT 1998. LNCS, vol. 1514, pp. 51–65. Springer, Heidelberg (1998)
13. Coron, J.S.: Resistance against differential power analysis for elliptic curve cryptosystems. In: Koç, Ç.K., Paar, C. (eds.) CHES 1999. LNCS, vol. 1717, pp. 292–302. Springer, Heidelberg (1999)

14. Fouque, P.A., Valette, F.: The Doubling Attack – *Why Upwards Is Better than Downwards*. In: Walter, C.D., Koç, Ç.K., Paar, C. (eds.) CHES 2003. LNCS, vol. 2779, pp. 269–280. Springer, Heidelberg (2003)

15. Homma, N., Miyamoto, A., Aoki, T., Satoh, A., Shamir, A.: Collision-based power analysis of modular exponentiation using chosen-message pairs. In: Oswald, E., Rohatgi, P. (eds.) CHES 2008. LNCS, vol. 5154, pp. 15–29. Springer, Heidelberg (2008)

16. Hutter, M., Schwabe, P.: NaCl on 8-Bit AVR microcontrollers. In: Youssef, A., Nitaj, A., Hassanien, A.E. (eds.) AFRICACRYPT 2013. LNCS, vol. 7918, pp. 156–172. Springer, Heidelberg (2013)

17. Blake, I.F., Seroussi, G., Smart, N.P.: Advances in Elliptic Curve Cryptography, vol. 317. Cambridge University Press, Cambridge (1999)

18. Riscure Inspector

19. Joye, M.: Elliptic curve cryptosystems and side channel analysis. ST J. Syst. Res. **4**, 17–21 (2003)

20. Joye, M., Tymen, C.: Protections against differential analysis for elliptic curve cryptography. In: Koç, Ç.K., Naccache, D., Paar, C. (eds.) CHES 2001. LNCS, vol. 2162, pp. 377–390. Springer, Heidelberg (2001)

21. Joye, M., Yen, S.-M.: The montgomery powering ladder. In: Kaliski Jr., B.S., Koç, Ç.K., Paar, C. (eds.) CHES 2002. LNCS, vol. 2523, pp. 291–302. Springer, Heidelberg (2003)

22. Koblitz, N.: Elliptic curve cryptosystems. Math. Comput. **48**, 203–209 (1987)

23. Kocher, P.C., Jaffe, J., Jun, B.: Differential power analysis. In: Wiener, M. (ed.) CRYPTO 1999. LNCS, vol. 1666, pp. 388–397. Springer, Heidelberg (1999)

24. ARM mbed. Polarssl version 1.3.7. https://tls.mbed.org/

25. ST Microelectronics: RM0090 Reference Manual. DocID018909 Rev 8 (2014)

26. Miller, V.S.: Use of elliptic curves in cryptography. In: Williams, H.C. (ed.) CRYPTO 1985. LNCS, vol. 218, pp. 417–426. Springer, Heidelberg (1986)

27. De Mulder, E., Buysschaert, P., Berna Örs, S., Delmotte, P., Preneel, B., Vandenbosch, G., Verbauwhede, I.: Electromagnetic analysis attack on an FPGA Implementation of an elliptic curve cryptosystem. In: IEEE International Conference on Computer as a Tool, Belgrade, Serbia & Montenegro, November 2005, pp. 1879–1882 (2005). doi:10.1109/EURCON.2005.1630348, http://www.sps.ele.tue.nl/members/m.j.bastiaans/spc/demulder.pdf

28. NIST: FIPS publication 186-4 - Digital Signature standard (DSS). Technical report, National Institute of Standards and Technology (NIST), July 2013

29. Rechberger, C., Oswald, E.: Practical template attacks. In: Lim, C.H., Yung, M. (eds.) WISA 2004. LNCS, vol. 3325, pp. 440–456. Springer, Heidelberg (2005)

30. Rivain, M.: Fast and regular algorithms for scalar multiplication over elliptic curves. IACR Cryptology ePrint Archive, 2011:338 (2011)

Fault Analysis

Algorithmic Countermeasures Against Fault Attacks and Power Analysis for RSA-CRT

Ágnes Kiss[1]([✉]), Juliane Krämer[1,2], Pablo Rauzy[3], and Jean-Pierre Seifert[2]

[1] TU Darmstadt, Darmstadt, Germany
agnes.kiss@crisp-da.de, jkraemer@cdc.informatik.tu-darmstadt.de
[2] TU Berlin, Berlin, Germany
{juliane,jpseifert}@sec.t-labs.tu-berlin.de
[3] Inria, CITI Lab, Villeurbanne, France
pablo.rauzy@inria.fr

Abstract. In this work, we analyze all existing RSA-CRT countermeasures against the Bellcore attack that use binary self-secure exponentiation algorithms. We test their security against a powerful adversary by simulating fault injections in a fault model that includes random, zeroing, and skipping faults at all possible fault locations. We find that most of the countermeasures are vulnerable and do not provide sufficient security against all attacks in this fault model. After investigating how additional measures can be included to counter all possible fault injections, we present three countermeasures which prevent both power analysis and many kinds of fault attacks.

Keywords: Bellcore attack · RSA-CRT · Modular exponentiation · Power analysis

1 Introduction

In a fault attack, an adversary is able to induce errors into the computation of a cryptographic algorithm and thereby to gain information about the secret key or other secret information used in the algorithm. The first fault attack [4] targets an RSA implementation using the Chinese remainder theorem, RSA-CRT, and is known as the Bellcore attack. The Bellcore attack aroused great interest and led to many publications about fault attacks on RSA-CRT, e.g., [1,6,9,11,22]. Countermeasures to prevent the Bellcore attack can be categorized into two families: the first one relies on a modification of the RSA modulus and the second one uses self-secure exponentiation. The countermeasures in the first family were recently analyzed [21], and a formal proof of their (in)security was provided.

We complement the work of [21] by comprehensively analyzing the countermeasures in the second family, i.e., those based on self-secure exponentiation. These countermeasures use specific algorithms that include redundancy within the exponentiations. The first such method is based on the Montgomery ladder [9]. This was adapted to the right-to-left version of the square-and-multiply-always algorithm [5,6] and to double exponentiation [18,22]. We test the security

© Springer International Publishing Switzerland 2016
F.-X. Standaert and E. Oswald (Eds.): COSADE 2016, LNCS 9689, pp. 111–129, 2016.
DOI: 10.1007/978-3-319-43283-0_7

of these methods using an automated testing framework. We use the same fault model as in [21], but extend it to meet the particularities of self-secure exponentiation algorithms. We reveal that the countermeasures have certain vulnerabilities in this extended fault model. Based on these findings, we improve the countermeasures and present three self-secure exponentiation methods that are secure against fault injections, safe-error attacks, and power analyses. We note that non-algorithmic level countermeasures are not in the scope of this paper.

Our Contribution: In this paper, we **test the security** of the self-secure exponentiation countermeasures **against the Bellcore attack** by simulating random, zeroing, and skipping faults at all possible fault locations (Sect. 4). Thereafter, we **propose secure countermeasures**, step-by-step achieving protection against all fault injections and resistance to power analysis and safe-error attacks. We present one countermeasure for each of the exponentiation algorithms used as self-secure exponentiation: the *Montgomery ladder*, the *square-and-multiply-always* algorithm, and the *double exponentiation* method. Despite the natural overhead caused by the included measures against all the considered attack types, **our algorithms remain highly efficient** (Sect. 5).

2   Background

In this section, we give the necessary background information for our work.

2.1   The Bellcore Attack on RSA-CRT

We use the standard notation for RSA [23]: M denotes the message, $N = pq$ the public modulus with secret primes p and q, $\varphi(N) = (p-1)(q-1)$. The public exponent e with $\gcd(e, \varphi(N)) = 1$ is chosen along with the secret exponent d, where $e \cdot d \equiv 1 \bmod \varphi(N)$. The signature is calculated $S = M^d \bmod N$, and $S^e \equiv (M^d)^e \equiv M \bmod N$. The calculation can be speeded up by a factor of four using the RSA-CRT implementation [20]. Two smaller exponentiations $S_p = M^{d_p} \bmod p$ and $S_q = M^{d_q} \bmod q$ are performed with exponents $d_p = d \bmod (p-1)$, $d_q = d \bmod (q-1)$, and recombined with the method $S = \mathrm{CRT}(S_p, S_q) = ((S_p - S_q)i_q \bmod p)q + S_q$, where $i_q = q^{-1} \bmod p$. The public key of RSA-CRT is (e, N) and the private key includes p, q, d_p, d_q and i_q.

A *fault attack* is a physical attack where the attacker is able to induce faults into the execution of the algorithm. The first attack on RSA-CRT was proposed by Bellcore researchers [4]. The fault is induced into the calculation of strictly one of the intermediate signatures, resulting in $\widehat{S_p}$ (or $\widehat{S_q}$). If $\widehat{S_p}$ (or $\widehat{S_q}$) is used during recombination, a faulty signature $\widehat{S}$ is returned. With high probability q (or p) can be deduced as $\gcd(S - \widehat{S}, N)$ [4] or as $\gcd(\widehat{S}^e - M \bmod N, N)$ [11].

During the discussion of fault attacks, the precise description of the *fault model* is essential: it includes the assumptions on the adversary's abilities. The Bellcore attack targeting an unprotected implementation uses one fault injection and loose assumptions in the fault model, i.e., a very weak attacker. The attacker is only assumed to alter an intermediate signature, which can be achieved by an arbitrary modification of any variable throughout the exponentiation, i.e., affecting any bit or any byte results in a successful attack.

2.2 Safe-Error Attacks

Classical fault attacks exploit the corrupted result or the difference between a correct and faulty results. However, it was noted in [26] that secret information may leak depending on if a fault has effect on the result of the computation or not. The techniques that exploit such behavior are called safe-error (SE) attacks.

Computational safe-error attacks (C-SE) [27] target dummy operations. If the result remains error-free although a fault was induced, it affects a dummy operation and thus, information about the secret key can be revealed.

Memory safe-error attacks (M-SE) [26] assume a more powerful attacker. Knowing how the internal variables are processed in the memory throughout a certain step of the algorithm, one may be able to derive the secret key [26]. Memory safe-error attacks are prevented by randomizing the targeted variables.

2.3 Power Analysis Methods

Simple power analysis (SPA) studies the power consumption of a single execution of the algorithm. If the execution depends on the value of the secret key, the adversary is able to obtain information by analyzing the power trace.

Differential power analysis (DPA) is a natural extension of SPA [16]. When performing a DPA, an attacker collects several power trace measurements of the executions of the same algorithm and uses statistical methods to reveal the secret key. Prevention generally requires randomization of variables.

2.4 Algorithms for Regular Exponentiation

Classical modular exponentiation algorithms are vulnerable to SPA, since the power consumption of the different operations can be differentiated [17]. To prevent SPA, regularity of the modular exponentiation algorithms is required. It means that the same operations are performed independently from the value of the exponent. Below, we recapitulate the two most widely used binary methods.

Algorithm 1. SPA-resistant modular exponentiation methods	
(1a) Square-and-multiply-always [7]	**(1b)** Montgomery ladder [13]
input: $M \neq 0$, $d = (d_{n-1}, \ldots, d_0)_2$, x **output:** $M^d \bmod x$	**input:** $M \neq 0$, $d = (d_{n-1}, \ldots, d_0)_2$, x **output:** $M^d \bmod x$
1: $R_0 := 1, R_1 := 1, R_2 := M$	1: $R_0 := 1, R_1 := M$
2: **for** $i = 0$ **to** $n - 1$ **do**	2: **for** $i = n - 1$ **to** 0 **do**
3: $R_{\overline{d_i}} := R_{\overline{d_i}} \cdot R_2 \bmod x$	3: $R_{\overline{d_i}} := R_{\overline{d_i}} \cdot R_{d_i} \bmod x$
4: $R_2 := R_2^2 \bmod x$	4: $R_{d_i} := R_{d_i}^2 \bmod x$
5: **end for**	5: **end for**
6: **return** R_0	6: **return** R_0

Square-and-Multiply-Always: The right-to-left exponentiation algorithm was modified in [7] to the square-and-multiply-always method, shown in Algorithm 1a. By introducing dummy operations in register R_1 (line 3), one squaring and one multiplication is performed at each iteration.

Montgomery Ladder: The powering ladder, shown in Algorithm 1b, was proposed in [19] and its correctness discussed in [13]. The algorithm is regular without including dummy operations and is resistant to safe-error attacks [13].

3   Countermeasures Against the Bellcore Attack

To counter the Bellcore attack, *straightforward countermeasures* aim to verify the integrity of the computation before returning the result, e.g., by repeating the computation and comparing the results. Due to the inefficiency of such measures, several improved countermeasures appeared starting from 1999.

3.1   Two Families of Countermeasures

The advanced countermeasures were divided into two families according to the difference in their nature [21]: *Shamir's family* and Giraud's family. We refer to the latter as *self-secure exponentiation countermeasures*.

Shamir's family consists of the countermeasures that prevent the Bellcore attack by multiplicatively extending the modulus x with a random number s. They rely on the fact that an invariant, inherited from the calculations modulo the extended modulus, i.e., modulo $x \cdot s$, must hold modulo s. Shamir's algorithm from [24] motivated researchers to develop such countermeasures, e.g., [1,12,21].

The idea of **self-secure exponentiation countermeasures** was proposed in [9]. If the exponentiation algorithm returns more than one power of a given input and keeps a *coherence* between its registers throughout the exponentiation, an invariant can be formulated that must hold at the end of the algorithm. However, it is claimed to be lost if a fault injection takes place.

3.2   Self-secure Exponentiation Countermeasures

In this section, we recapitulate the existing self-secure exponentiation countermeasures. The algorithms are provided in Appendix A in Algorithms 5–10.

The first countermeasure was proposed by **Giraud** [9]. It exploits the fact that while using the *Montgomery ladder*, the temporary registers R_0 and R_1 are of the form $(M^{k-1} \bmod x, M^k \bmod x)$ for some integer k after each iteration of Algorithm 1b. After two exponentiations that result in the pairs $(S'_p = M^{d_p-1} \bmod p, S_p = M^{d_p} \bmod p)$ and $(S'_q = M^{d_q-1} \bmod q, S_q = M^{d_q} \bmod q)$, and two recombinations $S' = \mathrm{CRT}(S'_p, S'_q) = M^{d-1} \bmod pq$ and $S = \mathrm{CRT}(S_p, S_q) = S^d \bmod pq$, the invariant $M \cdot S' \equiv S \bmod pq$ holds. Giraud claims that in case of a fault attack within the exponentiation, the coherence is lost for S_p, S'_p (or S_q, S'_q) and thus for S and S'. Despite its advantages, the Montgomery ladder exponentiation remains vulnerable to DPA [16] (DPA^{exp}).

Fumaroli and Vigilant blinded the base element with a small random number r [8], using one more register R_2 in the exponentiation. Besides being more memory-costly, this method was proven to be insecure against fault attacks [14], due to the lack of coherence between R_2 and the other registers. Moreover, it remains vulnerable to the DPA attack on the CRT recombination from [25] (DPA^{CRT}).

The *square-and-multiply-always algorithm* (Algorithm 1a), uses dummy operations to prevent SPA. **Boscher et al.** in 2007 proposed a self-secure exponentiation countermeasure based on this algorithm [6]. In the end of the execution, R_0 holds the value $M^d \bmod x$, R_1 holds $M^{2^n-d-1} \bmod x$, while R_2 only depends on the binary length n of the exponent, and equals to $M^{2^n} \bmod x$. Thus, the coherence $M \cdot R_0 \cdot R_1 \equiv R_2 \bmod x$ is kept throughout the algorithm. Boscher et al. in 2009 [5], modified the method in order to achieve resistance against DPA on the exponentiation without significant overhead. 2^w-ary versions of the algorithm were proposed [2,10].

Rivain proposed a solution that uses *double exponentiation* [22]. Such a method receives the base M, two exponents d_1, d_2, the modulus x, and outputs both $M^{d_1} \bmod x$ and $M^{d_2} \bmod x$. It makes use of a double addition chain for the pair (d_1, d_2), by means of which the two modular exponentiations are performed at once, using altogether $1.65n$ operations on average We assume this chain to be precomputed. **Le et al.** presented a double exponentiation algorithm, that does not rely on precomputation [18]. The binary exponentiation works as two parallel executions of the right-to-left exponentiation and uses register R_0 for calculations with d_1 and register R_1 for calculations with d_2. $M^{2^n} \bmod x$ is computed only once and is stored in R_2.

Table 1 summarizes the different properties of the self-secure exponentiation countermeasures. We consider the security and efficiency of the methods, since

Table 1. Self-secure exponentiation countermeasures. CRT, check, inv., reg., mult., and sq. denote the number of CRT recombinations, checking procedures, inversions, additional large registers, multiplications, and squaring operations respectively, in terms of the bit-length n of the exponent. PA and SE denote the resistance against power analysis and safe-error attacks. ✓ means that there are included countermeasures, × refers to the lack of them.

Countermeasure		Efficiency criteria						Physical attacks				
Author(s)	Ref.	CRT	Check	Inv.	Reg.	Mult.	Sq.	PA				SE
	Ref.	Alg.	Total			Per exp.			SPA	exp DPA	CRT DPA	C	M
Giraud	[9]	5	2	4	0	3	n	n	✓	×	✓	✓	✓
Fumaroli,Vigilant	[8]	6	2	4	$2^{(p,q)}$	4	$n+3$	$2n$	✓	✓	×	✓	✓
Boscher et al. '07	[6]	7	3	5	0	4	n	n	✓	×	×	✓	×
Boscher et al. '09	[5]	7	3	5	$1^{(pq)}$	4	$n+2$	n	✓	✓	×	✓	✓
Rivain	[22]	8	1	2	0	2	$1.65n$		×	×	×	✓	×
Rivain (SCA)	[22]	9	1	2	0	3	$1.65n$	0	✓	✓	×	✓	✓
Le et al.	[18]	10	1	2	0	3	$0.67n$	n	×	×	×	✓	×

measures against physical attacks imply overhead. When discussing efficiency, we describe the following relevant properties to achieve low time and memory consumption: number of registers containing large values that are used additionally to the input *registers* (M, d, x) during the exponentiation, number of *multiplications*, *squaring operations* and *inversions* using large registers. We summarize if they include protection against physical attacks such as *power analysis* on the exponentiation and the CRT recombination and *safe-error attacks*.

4 Security of Self-secure Exponentiation Methods

The security of self-secure exponentiation countermeasures relies mainly on the *exponentiation algorithms*. Each method has an invariant that holds throughout its execution, which is claimed to be lost in case a fault is injected. Accordingly, the modular exponentiation methods have to be tested against fault attacks. In this section, we recapitulate the fault model that we adopt, briefly describe our methodology and discuss our results.

4.1 Simulating Fault Injections Against Self-secure Exponentiation Countermeasures

The designers of the countermeasures provide either formal and informal explanations for their security assumptions and their fault models differ from each other. To the best of our knowledge, we are the first to simulate all possible fault injections in a common fault model.

Fault Model: We adopt the generic fault model of [21]. Therefore, we simulate three types of fault injections: *random* and *zeroing faults* in case of which the affected variable is changed to a random value and null, respectively, and *skipping faults* which cause instruction skips, i.e., jumps over some lines of the pseudocode. We take the following fault types into consideration: faults on local variables, on input parameters, and on conditional tests. An adversary is able to target any variable, but cannot specify the bits his fault affects. When inducing a random fault, he does not know its concrete value. Since refined methods appear for performing instruction skips in practice (e.g. [3]), we consider it as a possible threat when discussing physical attacks. The injection of skipping faults was observed as practical in [21], but was covered by means of random and zeroing faults. This does not hold for self-secure exponentiation. When considering skipping faults, we count the number of lines that have to be skipped in the pseudocode. In the Montgomery ladder shown in Algorithm 1b, the pair (R_0, R_1) is of the form $(M^{k-1} \bmod x, M^k \bmod x)$ at each iteration, which coherence is assumed to be lost in case of a fault injection. However, an adversary might skip two consecutive lines (3 and 4) at any iteration of the loop. The invariant holds for the corrupted $\widehat{R_0}$ and $\widehat{R_1}$ and thus, the fault is not detected.

Our Framework: In case of self-secure exponentiation countermeasures, the underlying *exponentiation algorithm* has to be tested and checked that the invariant is lost if a fault is injected. When simulating the attacks, we needed features

Table 2. Results of our fault injection tests on the exponentiation algorithms, assuming that the checking procedures are protected We note that we rely on the original fault models of the countermeasures from column Ref., recapitulated in Appendix A. $\checkmark$ denotes that our tests did not reveal any vulnerability against the fault type, M and d_1, d_2 denote the vulnerability of the message and the exponents in the exponentiation algorithm, respectively. When considering skipping faults, we indicate which lines are skipped together to achieve a successful attack. The register numbering R_i, $i \in \{0, 1, 2\}$ and the lines are according to the algorithms in column Alg.

Countermeasure			Fault injection attacks				
Author(s)	Ref.	Alg.	Random	Zeroing		Skipping	
Fault number			1	1	2	1	2
Giraud	[9]	5	$\checkmark$	M, R_0, R_1		$\checkmark$	(4–5)
Fumaroli, Vigilant	[8]	6	R_2	M, R_0, R_1, R_2		(7)	(5–6) or $2 \cdot$ (7)
Boscher et al. 2007	[6]	7	$\checkmark$	$\checkmark$	$\checkmark$	$\checkmark$	(6–7)
Boscher et al. 2009	[5]	7	$\checkmark$	$\checkmark$	$\checkmark$	$\checkmark$	(6–7)
Rivain	[22]	8	M	$\checkmark$	$\checkmark$	$\checkmark$	$\checkmark$
Rivain (SCA)	[22]	9	M	$\checkmark$	$\checkmark$	$\checkmark$	$\checkmark$
Le et al.	[18]	10	M	$\checkmark$	d_1, d_2	$\checkmark$	$\checkmark$

that the tool used for the analysis of Shamir's family lacked [21]: redefinition of variables and support for loops. Therefore, we created our own framework in Java. A manual step of our method was identifying the possible fault injection locations within the exponentiation algorithms. After this manual step, the simulation of multiple fault injections in all possible combination of fault locations was fully automated, for all the three fault types. A simulation results in a successful Bellcore attack if a corrupted signature is returned. For more details on our simulation framework, the reader is referred to the full version [15].

4.2   Simulation Results

The results of our fault injection simulations are shown in Table 2. While performing the tests with multiple faults, we considered protected checking procedures, since skipping or zeroing any of the checks would enable a successful Bellcore attack. When considering faults on the checking procedures, a method can be protected against n fault injections by repeating each check n times.

Random Faults: If a countermeasure is protected against one random fault injection, it cannot be broken with more than one random faults either. This is due to the fact that a random fault cannot induce a verification skip [21]. Our results confirm that in case of the algorithms that use the *Montgomery ladder* or the *square-and-multiply-always algorithm*, the intermediate secret exponent and the loop counter have to be protected against random faults. [6,8,9] use the *checksum* of the exponent to verify its integrity and thwart the attack. It was revealed in [14], that the introduction of **register R_2** in Fumaroli and Vigilant's countermeasure [8] made it vulnerable to any random fault on R_2 at any iteration of the algorithm. This is due to the fact that R_2 is calculated independently of the other two registers, which are multiplied with its final value. In case of the countermeasures using *double exponentiation*, a possible random fault is the corruption of the intermediate **message M**, resulting in $\widehat{M}$. Rivain identified this vulnerability and suggested to compute a cyclic redundancy code [22].

Zeroing Faults: Without a specific checking procedure against zeroing faults, the exponentiation algorithms (Sect. 2.4) are vulnerable. According to [9], it is unlikely to be able to zero a large buffer in practice. However, as [6,21], we take zeroing faults into consideration but note that their injection is very difficult to achieve in practice. In case of the methods that use the *Montgomery ladder* and the *square-and-multiply-always exponentiation*, if the **message M** in the beginning of the algorithms is zeroed, zeroes are returned. The same holds for any of the **registers R_0, R_1** in the method using the Montgomery ladder and for **R_2** in Fumaroli and Vigilant's and Boscher et al.'s methods. Then, the checking procedure holds even though the recombination is computed with only one of the intermediate signatures. Giraud considered this vulnerability impossible, while Boscher et al. included checks against it. The two countermeasures that use *double exponentiation* are not vulnerable to a single zeroing fault. In the case of Rivain's method [22], the exponent is given by the addition chain, which we assume to be protected. For the algorithm by Le et al. [18], two zeroing faults on the **exponents d_1, d_2** are necessary to conduct a Bellcore attack. If any other values are zeroed, the coherence check does not hold and the fault is detected.

Skipping Faults: Our simulations show that only the method by Fumaroli and Vigilant [8] is vulnerable to the instruction skip of **one line**, the calculation of register R_2, which has a similar effect as the random fault on R_2. When two lines are skipped together, both regular, SPA-resistant algorithms, i.e., the *Montgomery ladder* and the *square-and-multiply-always* methods are vulnerable. By skipping **two consecutive lines within the loop**, they preserve the coherence between the variables even though the results are corrupted. Even if the loop counter i is protected, skipping faults result in successful Bellcore attacks.

5   PA-SE-FA-Resistant Self-secure Exponentiation Countermeasures

We propose a secure countermeasure for each of the exponentiation algorithms that are used for constructing self-secure exponentiation methods We claim that our proposed countermeasures are secure against *power analysis* (PA), *safe-error* (SE) attacks, and *fault attacks* (FA) and remain highly efficient. For the verification of the resistance against fault injection attacks, we applied our framework from Sect. 4.1 on the proposed algorithms. We discuss the implied overhead by the introduced protection against physical attacks. FA_i^j denotes fault attacks of type j (r, z, s denote random, zeroing and skipping faults, resp.), against variable(s) i.

Algorithm 2. PA-SE-FA method with the Montgomery ladder

(2a) $\text{MONEXP}(M, d, x, r, r_{\text{inv}}, s)$

input: $M, d = (d_{n-1}, \ldots, d_0)_2,$
$x,\ r,\ r_{\text{inv}},\ s$
output: $(r^{2^n} \cdot M^d \bmod sx,$
$r^{2^n} \cdot M^{d+1} \bmod sx,$
$r_{\text{inv}}^{2^n} \bmod sx)$

1: $x := s \cdot x$ $\quad \triangleright \text{FA}^s_{(6-7)},\ \text{FA}^{r,\,z}_{d,i}$
2: $R_0 := r$
3: $R_1 := r \cdot M \bmod x$
4: $R_2 := r_{\text{inv}} \bmod x$
5: **for** i from $n-1$ to 0 **do**
6: $\quad R_{\overline{d_i}} := R_{\overline{d_i}} \cdot R_{d_i} \bmod x$
7: $\quad R_{d_i} := R_{d_i}^2 \bmod x$
8: $\quad R_2 := R_2^2 \bmod x$
9: **end for**
10: **return** (R_0, R_1, R_2)

(2b) RSA-CRT

input: $M \neq 0, p, q, d_p, d_q, i_q,$
$D = p \oplus q \oplus d_p \oplus d_q \oplus i_q$
output: $M^d \bmod pq$ or error

1: Pick k-bit random prime s,
$\quad$ such that $ps \nmid M, qs \nmid M$ $\quad \triangleright \text{FA}^s_{(6-7)},\ \text{FA}^{r,\,z}_{d,i}$
2: Pick random integer $r \in \mathbb{Z}^*_{pqs}$ $\quad \triangleright \text{FA}^r_{R_2},\ \text{FA}^s_{(8)}$
3: $r_{\text{inv}} := r^{-1} \bmod pqs$ $\quad \triangleright \text{FA}^r_{R_2},\ \text{FA}^s_{(8)}$
4: $(S_p, S'_p, R_p) := \text{MONEXP}(M \bmod sp, d_p, p, r, r_{\text{inv}}, s)$
5: $(S_q, S'_q, R_q) := \text{MONEXP}(M \bmod sq, d_q, q, r, r_{\text{inv}}, s)$
6: **if** $S_p \cdot S_q = 0$ **then** $\quad \triangleright \text{FA}^z_{M, R_0, R_1, R_2}$
7: $\quad$ **return** error
8: **end if**
9: $S := \text{CRT}_{\text{blinded}}(S_p, S_q)$ $\quad \triangleright \text{DPA}_{\text{CRT}}$
10: $S' := \text{CRT}_{\text{blinded}}(S'_p, S'_q)$ $\quad \triangleright \text{DPA}_{\text{CRT}}$
11: $R := \text{CRT}_{\text{blinded}}(R_p, R_q)$ $\quad \triangleright \text{FA}^r_{R_2},\ \text{FA}^s_{(8)}$
12: $S := R \cdot S \bmod pq$ $\quad \triangleright \text{FA}^r_{R_2},\ \text{FA}^s_{(8)}$
13: **if** $M \cdot S \not\equiv R \cdot S' \bmod pq$ **then**
14: $\quad$ **return** error
15: **end if**
16: $S_{ps} = (S_p \bmod s)^{d_q \bmod (s-1)} \bmod s$
17: $S_{qs} = (S_q \bmod s)^{d_p \bmod (s-1)} \bmod s$
18: **if** $S_{ps} \neq S_{qs}$ **then**
19: $\quad$ **return** error $\quad \triangleright \text{FA}^s_{(6-7)},\ \text{FA}^{r,\,z}_{d,i}$
20: **end if**
21: **if** $p \oplus q \oplus d_p \oplus d_q \oplus i_q \neq D$ **then**
22: $\quad$ **return** error $\quad \triangleright \text{FA}^{r,\,z}_{p, q, i_q, d_p, d_q}$
23: **end if**
24: **return** S

5.1 Countermeasure Using the Montgomery Ladder

Fumaroli and Vigilant's countermeasure [8] (Algorithm 6) which aimed to improve Giraud's method [9] (Algorithm 5) was proven to be vulnerable to random fault attacks [14]. Algorithm 2 presents our secure method with the *Montgomery ladder*.

To prevent fault attacks on **register R₂** (**$\text{FA}^r_{R_2}$, $\text{FA}^s_{(8)}$**), we return the blinded registers R_0 and R_1 and perform the multiplication with the inverse contained in R_2. This multiplication happens modulo pq, after the blinded CRT recombinations of all the three registers in lines 9–11 in Algorithm 2b.

To achieve prevention against **skipping faults** (**$\text{FA}^s_{(6-7)}$**), we include a check for verifying the integrity of the exponentiations. Since the coherence in the regular exponentiation algorithms is not lost when skipping faults are injected, we create a hybrid countermeasure with a technique used in Shamir's family by Aumüller et al. [1]. We conclude the necessity of the modulus extension to prevent skipping faults and multiply the modulus with a k-bit random prime s. S_p and S_q are calculated modulo $p \cdot s$ and $q \cdot s$, respectively, and the signature is recombined to $S = M^d \bmod pq$ using the blinded recombination from [9]:

$$S = \text{CRT}_{\text{blinded}}(S_p, S_q) = (((S_p - S_q) \bmod sp) \cdot i_q \bmod sp) \cdot q + S_q \bmod pq. \quad (1)$$

To verify that no instruction was skipped, two small exponentiations modulo the k-bit number s with the k-bit exponents are performed as in lines 16 and 17. If a skipping fault occurs and the value of S_p or S_q is corrupted, the check in line 18 does not hold with probability 2^{-k}. Besides protecting against skipping faults, this measure detects faults on the **exponent** and **loop counter i** ($\mathbf{FA_{d,i}^{r,z}}$) of the exponentiation algorithm, without an additional large register. If the small exponentiations are calculated using the Montgomery ladder (Algorithm 1b), then besides the k-bit message, exponent, and modulus, two k-bit registers, k multiplications and squarings are used. However, a checksum as an input has to be included to detect the corruption of p, q, i_q, d_p or d_q in Algorithm 2b in line 21.

We note that the blinded CRT recombination recapitulated in Eq. 1 also prevents the **DPA** attack on the CRT recombination (**DPA$_{\text{CRT}}$**) from [25].

To avoid **zeroing faults** ($\mathbf{FA_{M,R_0,R_1,R_2}^z}$), we check that none of the values returned by the exponentiation is zero. We perform this before the CRT recombinations in Algorithm 2b, by verifying $S_p \cdot S_q \neq 0$ in line 6. In order to make sure that this check does not violate the correctness of the algorithm when the message is a multiple of ps or qs, we choose s such that $ps \nmid M$ and $qs \nmid M$.

Algorithm 2 presents the algorithm that is based on the Montgomery ladder and is protected against power analysis (PA), safe-error (SE), and fault attacks (FA). For eliminating the revealed vulnerabilities against fault injection attacks, we included an additional CRT recombination, transformed two small inversions to one of doubled size, included one large input register D, two times k multiplications and k squaring operations on k-bit registers, where k is the security parameter that defines the probability of undetected skipping faults as 2^{-k}. We note that since modular inversion and prime generation imply significant costs, lines (1–3) can be precomputed (without the assumption $ps \nmid M, qs \nmid M$) and s, r and r_{inv} can be provided as inputs to Algorithm 2b.

5.2   Countermeasure Using the Square-and-Multiply-Always Exp.

Boscher et al. described a *square-and-multiply always algorithm* that is resistant to SPA, DPA, and SE [5] (Algorithm 7). The algorithm includes a technique against the exponent modification, and the check $R_2 \neq 0$ in the end of the exponentiation to detect **zeroing faults** ($\mathbf{FA_{M,R_2}^z}$) [6]. Instead of this check in both exponentiations, we suggest to verify $S_p \cdot S_q \neq 0$ in Algorithm 3b as in Algorithm 2b.

Algorithm 3. PA-SE-FA method with the square-and-multiply-always exp.

(3a) $\text{SQEXP}(M, d, x, r, r_{\text{inv}}, s)$	(3b) RSA-CRT

(3a) $\text{SQEXP}(M, d, x, r, r_{\text{inv}}, s)$

    input: $M, d = (d_{n-1}, \ldots, d_0)_2,$
      x, r, r_{inv}, s
    output: $(r \cdot M^d \bmod sx,$
      $r_{\text{inv}} \cdot M^{2^n - d - 1} \bmod sx,$
      $M^{2^n} \bmod sx)$

  1: $x := s \cdot x$ $\triangleright \text{FA}^s_{(6-7)}, \text{FA}^{r,z}_{d,i}$
  2: $R_0 := r$
  3: $R_1 := r_{\text{inv}}$
  4: $R_2 := M$

  5: **for** i from 0 to $n - 1$ **do**
  6:    $R_{\overline{d_i}} := R_{\overline{d_i}} \cdot R_2 \bmod x$
  7:    $R_2 := R_2^2 \bmod x$
  8: **end for**

  9: **return** (R_0, R_1, R_2)

(3b) RSA-CRT

    input: $M \neq 0, p, q, d_p, d_q, i_q,$
      $D = p \oplus q \oplus d_p \oplus d_q \oplus i_q$
    output: $M^d \bmod pq$ or error

  1: Pick k-bit random prime s
     such that $ps \nmid M, qs \nmid M$ $\triangleright \text{FA}^s_{(6-7)}, \text{FA}^{r,z}_{d,i}$
  2: Pick random integer $r \in \mathbb{Z}^*_{pqs}$ $\triangleright \text{FA}^r_{R_2}, \text{FA}^s_{(8)}$
  3: $r_{\text{inv}} := r^{-1} \bmod pqs$
  4: $(S_p, S'_p, T_p) := \text{SQEXP}(M \bmod sp, d_p, p, r, r_{\text{inv}}, s)$
  5: $(S_q, S'_q, T_q) := \text{SQEXP}(M \bmod sq, d_q, q, r, r_{\text{inv}}, s)$
  6: **if** $S_p \cdot S_q = 0$ **then** $\triangleright \text{FA}^z_{M, R_2}$
  7:    **return** error
  8: **end if**

  9: $S := \text{CRT}_{\text{blinded}}(S_p, S_q)$
10: $S' := \text{CRT}_{\text{blinded}}(S'_p, S'_q)$
11: $T := \text{CRT}_{\text{blinded}}(T_p, T_q)$
12: **if** $M \cdot S \cdot S' \not\equiv T \bmod pq$ **then**
13:    **return** error
14: **end if**

15: $S_{ps} = (r_{\text{inv}} S_p \bmod s)^{d_q \bmod (s-1)} \bmod s$
16: $S_{qs} = (r_{\text{inv}} S_q \bmod s)^{d_p \bmod (s-1)} \bmod s$
17: **if** $S_{ps} \neq S_{qs}$ **then**
18:    **return** error $\triangleright \text{FA}^s_{(6-7)}, \text{FA}^{r,z}_{d,i}$
19: **end if**

20: **if** $p \oplus q \oplus d_p \oplus d_q \oplus i_q \neq D$ **then**
21:    **return** error $\triangleright \text{FA}^{r, z}_{p, q, i_q, d_p, d_q}$
22: **end if**
23: **return** $r_{\text{inv}} \cdot S \bmod pq$

Against **skipping faults** (FA^s_{6-7}) we suggest the same measure as in Algorithm 2: blinding the modulus and performing two small exponentiations in the RSA-CRT algorithm. For retrieving the signature, the CRT recombination in Eq. 1 is used. Though not mentioned in [5], the random value r in Algorithm 3b should not be too small to avoid the following SPA during the computation of Algorithm 3a: if an adversary is allowed to input the message $M = 1$, the value of register R_2 remains 1 for the whole computation. Therefore, the multiplication in line 6 would only depend on the bits of the secret exponent d, multiplied either with a small number (r) or with a large number (r_{inv}). This could result in differences in the power consumption trace and therefore we chose r to be an at least $(n + k)$-bit integer, where n is the bitlength of p and of q, since it is used for operations of that size in Algorithm 3a.

Algorithm 4. PA-SE-FA method with double exponentiation

(4a) DOUBLEEXP(M, d_1, d_2, x, s)	(4b) RSA-CRT

(4a) DOUBLEEXP(M, d_1, d_2, x, s)

input: $M \neq 0$,
    $d_1 = (d_{1,n-1}, \ldots, d_{1,0})_2$,
    $d_2 = (d_{2,n-1}, \ldots, d_{2,0})_2, x, s$
output: $(M^{d_1} \bmod xs,$
    $M^{d_2} \bmod xs)$

1: $x := s \cdot x$ ▷ DPA$_{\text{CRT}}$
2: $R_{(0,1)} := 1$ ▷ SPA
3: $R_{(1,1)} := 1$ ▷ SPA
4: $R_{(0,2)} := 1$ ▷ SPA
5: $R_{(1,2)} := 1$ ▷ SPA
6: $R_2 := M$

7: **for** $i = 0$ **to** $n - 1$ **do**   ▷ SPA
8:      $R_{(\overline{d_{1,i}},1)} := R_{(\overline{d_{1,i}},1)} \cdot R_2 \bmod x$
9:      $R_{(\overline{d_{2,i}},2)} := R_{(\overline{d_{2,i}},2)} \cdot R_2 \bmod x$
10:      $R_2 := R_2^2 \bmod x$
11: **end for**
12: **if** $R_{(0,1)}R_{(1,1)} \not\equiv R_{(0,2)}R_{(1,2)} \bmod x$
     then   ▷ C SE
13:      **return** error
14: **end if**
15: **return** $(R_{(0,1)}, R_{(0,2)})$

(4b) RSA-CRT

input: M, p, q, d_p, d_q, i_q
output: $M^d \bmod pq$ or error

1: Pick small $r_1, r_2 \in \mathbb{Z}$ $r_2 \geq r_1 + 2$
2: Pick k-bit random prime s
3: $(S_p, c_p) :=$ ▷ DPA, M-SE, FA$_M^r$, FA$_{(d_1,d_2)}^z$
     DOUBLEEXP$(M \bmod p, d_p + r_1(p-1),$
     $r_2(p-1) - d_p - 1, p, s)$
4: $(S_q, c_q) :=$ ▷ DPA, M-SE, FA$_M^r$, FA$_{(d_1,d_2)}^z$
     DOUBLEEXP$(M \bmod q, d_q + r_1(q-1),$
     $r_2(q-1) - d_q - 1, q, s)$
5: $S := \text{CRT}_{\text{blinded}}(S_p, S_q)$ ▷ DPA$_{\text{CRT}}$
6: **if** $M \cdot S \cdot c_p \not\equiv 1 \bmod p$ **then**
7:      **return** error   ▷ FA$_M^r$, FA$_{(d_1,d_2)}^z$
8: **end if**
9: **if** $M \cdot S \cdot c_q \not\equiv 1 \bmod q$ **then**
10:      **return** error   ▷ FA$_M^r$, FA$_{(d_1,d_2)}^z$
11: **end if**
12: **return** $S \bmod pq$

Our PA-SE-FA-resistant algorithm with the square-and-multiply-always exponentiation is depicted in Algorithm 3. To eliminate the identified vulnerabilities, we included one large input register D along with two times k multiplications and k squaring operations on k-bit registers, in a similar manner as in Algorithm 2.

5.3 Countermeasure Using Double Exponentiation

Rivain proposed the first countermeasure that uses **double exponentiation** [22] (Algorithm 8). He included modifications by means of which it becomes SPA-DPA-SE-resistant, still requiring the precomputation of the addition chain (Algorithm 9). Our aim is to consider measures in the insecure but more efficient algorithm by Le et al. [18] (Algorithm 10), which does not include precomputation but ignores protection against PA and SE.

Firstly, we transform the algorithm to become resistant to **SPA**. We use two additional registers with dummy operations in order to achieve regularity. Thus, the algorithm requires the use of altogether 5 registers: $R_{(0,1)}$ and $R_{(1,1)}$ belonging to exponent d_1, $R_{(0,2)}$ and $R_{(1,2)}$ belonging to exponent d_2, and R_2 used as before. Since for every bit of the exponents the same operations have to performed, this results in altogether $2n$ multiplications and n squaring operations.

Introducing regularity includes dummy operations. Registers $R_{(1,1)}$ and $R_{(1,2)}$ are unused and thus all the multiplications that assign values to them are dummy operations. To avoid **computational safe-error attacks (C-SE)** on these operations, in the end of the exponentiation we include the check whether $R_{(0,1)} \cdot R_{(1,1)} \equiv R_{(0,2)} \cdot R_{(1,2)} \bmod x$. Since both the products corresponding to the two exponents are $M^{2^n-1} \bmod x$, this holds if the values are not corrupted. With this, we verify the correctness of the dummy values.

To achieve resistance against **differential power analysis** on the exponentiation ($\mathbf{DPA_{exp}}$) and **memory safe-error attacks (M-SE)**, we include the exponent blinding method of Rivain in the RSA-CRT algorithm [22]. Against DPA on the CRT recombination ($\mathbf{DPA_{CRT}}$), we apply the blinded CRT recombination method with extended modulus from [9]. For the description of r_1 and r_2 and the correctness of the blinding method, the reader is referred to [9,22].

To detect any randomizing fault on the **message M ($\mathbf{FA_M^r}$)**, we include its value in the coherence checks as it was seen in case of the countermeasures from [5,6,8,9]. We decrease the value of the exponents used for the calculation of c_p and c_q by one, and multiply the results with M, during the verification in lines 7 and 10 of Algorithm 4b. For instance, if S_p and c_p are calculated by means of a corrupted $\widehat{M}$, the verification $M \cdot \widehat{M}^{d_p+r_1\varphi(p)} \cdot \widehat{M}^{r_2\varphi(p)-d_p-1} \equiv 1 \bmod p$ does not hold with high probability. With this, the zeroing faults on **exponents $\mathbf{d_1}$ and $\mathbf{d_2}$ ($\mathbf{FA_{(d_1,d_2)}^z}$)** are also thwarted, the algorithm returns $(1,1)$ in case of two null exponents, and the modified check does not hold anymore.

Our PA-SE-FA-resistant countermeasure using double exponentiation is depicted in Algorithm 4. Though the modified countermeasure is less memory-efficient than Le et al.'s algorithm, we note its advantage against physical attacks.

Table 3. Comparison of our PA-SE-FA self-secure exponentiation countermeasures with previous methods. The notation is consistent with that of Tables 1 and 2, k denoting the included k operations (squaring and multiplication). We highlight with bold checkmarks (✓) those vulnerabilities that we eliminated in our secure countermeasures and we bold the additional resources needed to be used in order to achieve security against all the considered attacks.

Method		Efficiency criteria								Fault injection attacks					Other	
Ref	Alg	CRT	Check	Inv	Reg	k	Reg	Mult.	Sq.	Ran	Zeroing		Skipping		PA	SE
		Total					Per exp.				1	2	1	2		
[8]	6	2	4	$2^{(p,q)}$	0	0	4	$n+3$	$2n$	R_2	$M, R_\forall$		(7)	(5–6),2(7)	✓	✓
	3	**3**	4	$\mathbf{1^{(pqs)}}$	**1**	**4k**	**3**	$n+2$	$2n$	✓	✓	✓	✓	✓	✓	✓
[5,6]	7	3	5	$1^{(pq)}$	0	0	4	$n+2$	n	✓	✓	✓	✓	(6–7)	✓	✓
	3	3	4	$1^{(pqs)}$	**1**	**4k**	**3**	$n+1$	n	✓	✓	✓	✓	✓	✓	✓
[22]	9	1	2	0	0	0	3	$1.65n$	0	M	✓	✓	✓	✓	✓	✓
[18]	10	1	2	0	0	0	3	$1.65n$		M	✓	d_1, d_2	✓	✓	✗	✗
	4	1	**4**	0	0	0	**5**	$\mathbf{2n+3}$	n	✓	✓	✓	✓	✓	✓	✓

6   Conclusion

In this paper, we analyzed the existing self-secure exponentiation countermeasures against the Bellcore attack on RSA-CRT. Using our framework, we simulated all possible fault injections considering random and zeroing faults as well as instruction skips on the lines of pseudocode. We found that all the countermeasures using regular exponentiation algorithms lacked protection against some kind of faults or power analyses.

We presented three countermeasures, one for each exponentiation algorithm used for designing self-secure exponentiation countermeasures (cf. Table 3). All the three methods are based on regular algorithms to prevent *simple power analysis* (SPA), include randomization to be resistant to *differential power analysis* (DPA) and *memory safe-error* (M-SE) attacks, and eliminate dummy operations which could be exploited by *computational safe-error* (C-SE) attacks. Measures are included against all considered *fault injection attacks* (FA) as well. We verified that we eliminated the previous vulnerabilities of the methods without introducing new ones by applying our simulation framework on the pseudocode of the improved algorithms. To prevent skipping faults, we included additional checks into two of our methods, inspired by a countermeasure in Shamir's family, resulting in hybrid methods. We included prevention against fault attacks on the previously vulnerable register in the countermeasure that uses the Montgomery ladder. Our proposed solution that uses double exponentiation includes protection against power analyses and safe-error attacks in the algorithm where it was not considered.

We note that the vulnerability of the message corruption and of the DPA on the CRT recombination in Rivain's SPA-resistant method can be eliminated in a similar algorithmic manner as in Sect. 5.3, gaining another, the most efficient secure software countermeasure when precomputation is allowed. When precomputation is not allowed, our proposed solution using the square-and-multiply-always algorithm is the most efficient algorithmic countermeasure.

Acknowledgments. This work has been co-funded by the DFG as part of projects P1 and S5 within the CRC 1119 CROSSING and by the European Union's 7th Framework Program (FP7/2007-2013) under grant agreement no. 609611 (PRACTICE).

A Self-secure Exponentiation Countermeasures

Algorithm 5. Giraud's countermeasure [9]
PA attack model: SPA, chosen message SPA from [28].
Fault model: Random faults on variables and input parameters. Zeroing attacks, disruption of checking are regarded as impossible in practice. For the integrity check of d, i, we assume that an additional register is used in Table 1.

(5a) Modular exp.: $\text{GIREXP}(M, d, x, r)$	**(5b)** Giraud's RSA-CRT
input: $M, d = (d_{n-1}, \ldots, d_0)_2$ odd, x, r **output:** $(M^{d-1} \bmod r \cdot x, M^d \bmod r \cdot x)$	**input:** M, p, q, d_p, d_q, i_q **output:** $M^d \bmod pq$ or error
1: $x_r := r \cdot x$	1: Pick k-bit random prime r
2: $R_0 := M$, $R_1 := R_0^2 \bmod x_r$	2: $(S_p', S_p) := \text{GIREXP}(M \bmod p, d_p, p, r)$
3: **for** i from $n-2$ to 1 **do**	3: $(S_q', S_q) := \text{GIREXP}(M \bmod q, d_q, q, r)$
4: $\quad R_{\overline{d_i}} := R_{\overline{d_i}} \cdot R_{d_i} \bmod x_r$	4: $S := \text{CRT}_{\text{blinded}}(S_p, S_q)$
5: $\quad R_{d_i} := R_{d_i}^2 \bmod x_r$	5: $S' := \text{CRT}_{\text{blinded}}(S_p', S_q')$
6: **end for**	6: $S' := M \cdot S' \bmod (p \cdot q)$
7: $R_1 := R_1 \cdot R_0 \bmod x_r$	7: **if** $S' \neq S$ **then return** error
8: $R_0 := R_0^2 \bmod x_r$	8: **end if**
9: **if** i or d disturbed **then**	9: **if** p, q or i_q disturbed **then**
10: $\quad$ **return** error	10: $\quad$ **return** error
11: **end if**	11: **end if**
12: **return** (R_0, R_1)	12: **return** S

Algorithm 6. Fumaroli and Vigilant's countermeasure [8]
Attack model: SPA, DPA, against which blinding is included.
Fault model: That of Giraud's [9].

(6a) Modular exp.: $\text{FUMVIGEXP}(M, d, x)$	**(6b)** Fumaroli and Vigilant's RSA-CRT
input: $M \neq 0, d = (d_{n-1}, \ldots, d_0)_2$, x **output:** $(M^d \bmod x, M^{d+1} \bmod x)$	**input:** $M \neq 0, p, q, d_p, d_q, i_q$ **output:** $M^d \bmod pq$ or error
1: Pick k-bit random prime r	1: $(S_p, S_p') := \text{FUMVIGEXP}(M \bmod p, d_p, p)$
2: $R_0 := r$, $R_1 := rM \bmod x$	2: $(S_q, S_q') := \text{FUMVIGEXP}(M \bmod q, d_q, q)$
3: $R_2 := r^{-1} \bmod x$, $D := 0$	3: $S := \text{CRT}(S_p, S_q)$
4: **for** i from $n-1$ to 0 **do**	4: $S' := \text{CRT}(S_p', S_q')$
5: $\quad R_{\overline{d_i}} := R_{\overline{d_i}} \cdot R_{d_i} \bmod x$	5: **if** $S \cdot M \bmod p \cdot q \not\equiv S'$ **then**
6: $\quad R_{d_i} := R_{d_i}^2 \bmod x$	6: $\quad$ **return** error
7: $\quad R_2 := R_2^2 \bmod x$	7: **end if**
8: $\quad D := D + d_i,$	8: **if** p, q or i_q disturbed **then**
9: $\quad D := D \cdot 2$	9: $\quad$ **return** error
10: **end for**	10: **end if**
11: $D := D/2$	11: **return** S
12: $R_2 := R_2 \oplus D \oplus d$	
13: **return** $(R_2 \cdot R_0 \bmod x, R_2 \cdot R_1 \bmod x)$	

Algorithm 7. Boscher et al.'s countermeasure 2007 [6], **modifications 2009** [5]
Attack model: Regularity against SPA, **blinding against DPA**.
Fault model: One fault per execution [6], on local variables, input parameters.

(7a) Modular exp: $\text{BosExp}(M,d,x,\mathbf{r},\mathbf{r_{inv}})$	(7b) Boscher et al.'s RSA-CRT
input: $M,d = (d_{n-1},\ldots,d_0)_2,x,\mathbf{r},\mathbf{r_{inv}}$ **output:** $(\mathbf{r} \cdot M^d \bmod x,$ $\mathbf{r_{inv}} \cdot M^{2^n - d - 1} \bmod x, M^{2^n} \bmod x)$	**input:** $M \neq 0, p, q, d_p, d_q, i_q$ **output:** $M^d \bmod pq$ or error
1: $R_0 := 1 \cdot \mathbf{r}$	1: **Pick a k-bit random integer r**
2: $R_1 := 1 \cdot \mathbf{r_{inv}}$	2: $\mathbf{r_{inv}} := \mathbf{r}^{-1} \bmod pq$
3: $R_2 := M$	3: $(S_p, S_p', T_p) :=$
4: $D := 0$	$\text{BosExp}(M \bmod p, d_p, p, \mathbf{r}, \mathbf{r_{inv}})$
5: **for** i from 0 to $n-1$ **do**	4: $(S_q, S_q', T_q) :=$
6: $R_{\overline{d_i}} := R_{\overline{d_i}} \cdot R_2 \bmod x$	$\text{BosExp}(M \bmod q, d_q, q, \mathbf{r}, \mathbf{r_{inv}})$
7: $R_2 := R_2^2 \bmod x$	5: $S := \text{CRT}(S_p, S_q)$
8: $D := D + 2^n \cdot d_i$	6: $S' := \text{CRT}(S_p', S_q')$
9: $D := D/2$	7: $T := \text{CRT}(T_p, T_q)$
10: **end for**	8: **if** $M \cdot S \cdot S' \not\equiv T \bmod pq$ **then**
11: **if** $(D \neq d)$ or $(R_2 = 0)$ **then**	9: **return** error
12: **return** error	10: **end if**
13: **end if**	11: **return** $\mathbf{r_{inv}} \cdot S \bmod pq$
14: **return** (R_0, R_1, R_2)	

Algorithm 8. Rivain's countermeasure [22]
The addition chain is precomputed with $\text{ChainCompute}(d_1, d_2)$ from [22] and stored in memory or is computed on-the-fly.

(8a) Double exp.: $\text{RivExp}(M,\omega(d_1,d_2),x)$	(8b) Rivain's RSA-CRT
input: $M, \omega(d_1,d_2)$ n-bits chain, $d_1 \leq$ d_2, x **output:** $(M^{d_1} \bmod x, M^{d_2} \bmod x)$	**input:** M, p, q, d_p, d_q, i_q **output:** $M^d \bmod pq$ or error
1: $R_0 := 1, R_1 := M, \gamma := 1, i := 1$	1: $\omega_p := \text{ChainCompute}(d_p, 2(p-1) - d_p)$
2: **for** $i = 1$ to n **do**	2: $(S_p, c_p) := \text{RivExp}(M \bmod p, \omega_p, p)$
3: **if** $(\omega_i = 0)$ **then**	3: $\omega_q := \text{ChainCompute}(d_q, 2(q-1) - d_q)$
4: $R_\gamma := R_\gamma^2 \bmod x$	4: $(S_q, c_q) := \text{RivExp}(M \bmod q, \omega_q, q)$
5: $i := i + 1$	5: $S := \text{CRT}(S_p, S_q)$
6: **if** $(\omega_i = 1)$ **then**	6: **if** $S \cdot c_p \not\equiv 1 \bmod p$ **then**
7: $R_\gamma := R_\gamma \cdot M \bmod x$	7: **return** error
8: **end if**	8: **end if**
9: **else**	9: **if** $S \cdot c_q \not\equiv 1 \bmod q$ **then**
10: $R_{\gamma \oplus 1} := R_{\gamma \oplus 1} \cdot R_\gamma \bmod x$	10: **return** error
11: $\gamma := \gamma \oplus 1$	11: **end if**
12: **end if**	12: **return** S
13: **end for**	
14: **return** $(R_{\gamma \oplus 1}, R_\gamma)$	

Algorithm 9. Rivain's PA-resistant countermeasure [22]
Attack model: Regular SECRIVEXP and CHAINCOM against SPA, blinding against DPA. This blinding method can only be used if the double addition chain is computed on-the-fly.
Fault model: M is assumed to be protected, transient faults, i.e., faults whose effect lasts for one computation, are considered.

(9a) Double exp:	(9b) RSA-CRT
SECRIVEXP($M,\omega(d_1,d_2),x$)	

<table>
<tr><td>

(9a) Double exp:
SECRIVEXP($M,\omega(d_1,d_2),x$)

    input: $M \neq 0$, $\omega(d_1,d_2)$ n-bits,
    $d_1 \leq d_2$, x
    output: $(M^{d_1} \bmod x, M^{d_2} \bmod x)$

1: $R_{(0,0)} := 1, R_{(0,1)} := M,$
2: $R_{(1,0)} := M$
3: $\gamma := 1, \mu := 1, i := 0$

4: **while** $i < n$ **do**
5:    $t := \omega_i \wedge \mu$
6:    $v := \omega_{i+1} \wedge \mu$
7:    $R_{(0,\gamma \oplus t)} :=$
      $R_{(0,\gamma \oplus t)} \cdot R_{((\mu \oplus 1),\gamma \wedge \mu)} \bmod x$
8:    $\mu := t \vee (v \oplus 1)$
9:    $\gamma := \gamma \oplus t$
10:   $i := i + \mu + \mu \wedge (t \oplus 1)$
11: **end while**

12: **return** $(R_{\gamma \oplus 1}, R_\gamma)$

</td><td>

(9b) RSA-CRT

    input: M, p, q, d_p, d_q, i_q
    output: $M^d \bmod pq$ or error

1: Pick small $r_1, r_2 \in \mathbb{Z}$ $r_2 \geq r_1 + 2$
2: $\omega_p :=$
   CHAINCOM($d_p + r_1(p-1), r_2(p-1) - d_p$)
3: $(S_p, c_p) := $ SECRIVEXP($M \bmod p, \omega_p, p$)

4: $\omega_q :=$
   CHAINCOM($d_q + r_1(q-1), r_2(q-1) - d_q$)
5: $(S_q, c_q) := $ SECRIVEXP($M \bmod q, \omega_q, q$)

6: $S := \text{CRT}(S_p, S_q)$

7: **if** $S \cdot c_p \not\equiv 1 \bmod p$ **then**
8:    **return** error
9: **end if**

10: **if** $S \cdot c_q \not\equiv 1 \bmod q$ **then**
11:    **return** error
12: **end if**

13: **return** $S \bmod pq$

</td></tr>
</table>

Algorithm 10. Le et al.'s binary countermeasure [18]
Attack model: No side-channel attacks are discussed in [18].
Fault model: Same as that of Rivain [22].

<table>
<tr><td>

(10a) Double exp.: LEEXP(M, d_1, d_2, x)

    input: $M \neq 0, d_1 = (d_{1,n-1}, \ldots, d_{1,0})_2$
    $d_2 = (d_{2,n-1}, \ldots, d_{2,0})_2, x$
    output: $(M^{d_1} \bmod x, M^{d_2} \bmod x)$

1: $R_0 := 1, R_1 := 1, R_2 := M$

2: **for** $i = 0$ **to** $n - 1$ **do**
3:    **if** $d_{1,i} = 1$ **then**
4:       $R_0 := R_0 \cdot R_2 \bmod x$
5:    **end if**
6:    **if** $d_{2,i} = 1$ **then**
7:       $R_1 := R_1 \cdot R_2 \bmod x$
8:    **end if**
9:    $R_2 := R_2^2 \bmod x$
10: **end for**

11: **return** (R_0, R_1)

</td><td>

(10b) Rivain's RSA-CRT

    input: $M \neq 0, p, q, d_p, d_q, i_q$
    output: $M^d \bmod pq$ or error

1: $(S_p, c_p) := $ LEEXP($M \bmod p$,
      $d_p, 2(p-1) - d_p, p$)
2: $(S_q, c_q) := $ LEEXP($M \bmod q$,
      $d_q, 2(q-1) - d_q, q$)
3: $S := \text{CRT}(S_p, S_q)$

4: **if** $S \cdot c_p \not\equiv 1 \bmod p$ **then**
5:    **return** error
6: **end if**

7: **if** $S \cdot c_q \not\equiv 1 \bmod q$ **then**
8:    **return** error
9: **end if**

10: **return** S

</td></tr>
</table>

References

1. Aumüller, C., Bier, P., Fischer, W., Hofreiter, P., Seifert, J.: Fault attacks on RSA with CRT: concrete results and practical countermeasures. In: Kaliski Jr., B.S., Koç, Ç.K., Paar, C. (eds.) CHES 2002. LNCS, vol. 2523, pp. 260–275. Springer, Heidelberg (2003)
2. Baek, Y.: Regular 2^w-ary right-to-left exponentiation algorithm with very efficient DPA and FA countermeasures. Int. J. Inf. Sec. **9**(5), 363–370 (2010)
3. Blömer, J., Gomes Da Silva, R., Gunther, P., Kramer, J., Seifert, J.P.: A practical second-order fault attack against a real-world pairing implementation. In: Fault Diagnosis and Tolerance in Cryptography (FDTC 2014), pp. 123–136. IEEE (2014)
4. Boneh, D., DeMillo, R.A., Lipton, R.J.: On the importance of checking cryptographic protocols for faults. In: Fumy, W. (ed.) EUROCRYPT 1997. LNCS, vol. 1233, pp. 37–51. Springer, Heidelberg (1997)
5. Boscher, A., Handschuh, H., Trichina, E.: Blinded fault resistant exponentiation-revisited. In: Fault Diagnosis and Tolerance in Cryptography (FDTC 2009), pp. 3–9.IEEE (2009)
6. Boscher, A., Naciri, R., Prouff, E.: CRT RSA algorithm protected against fault attacks. In: Sauveron, D., Markantonakis, K., Bilas, A., Quisquater, J.-J. (eds.) WISTP 2007. LNCS, vol. 4462, pp. 229–243. Springer, Heidelberg (2007)
7. Coron, J.-S.: Resistance against differential power analysis for elliptic curve cryptosystems. In: Koç, Ç.K., Paar, C. (eds.) CHES 1999. LNCS, vol. 1717, pp. 292–302. Springer, Heidelberg (1999)
8. Fumaroli, G., Vigilant, D.: Blinded fault resistant exponentiation. In: Breveglieri, L., Koren, I., Naccache, D., Seifert, J.-P. (eds.) FDTC 2006. LNCS, vol. 4236, pp. 62–70. Springer, Heidelberg (2006)
9. Giraud, C.: An RSA implementation resistant to fault attacks and to simple power analysis. IEEE Trans. Comput. **55**(9), 1116–1120 (2006)
10. Joye, M., Karroumi, M.: Memory-efficient fault countermeasures. In: Prouff, E. (ed.) CARDIS 2011. LNCS, vol. 7079, pp. 84–101. Springer, Heidelberg (2011)
11. Joye, M., Lenstra, A.K., Quisquater, J.: Chinese remaindering based cryptosystems in the presence of faults. J. Cryptol. **12**(4), 241–245 (1999)
12. Joye, M., Paillier, P., Yen, S.M.: Secure evaluation of modular functions. In: 2001 International Workshop on Cryptology and Network Security (2001)
13. Joye, M., Yen, S.: The Montgomery powering ladder. In: Kaliski Jr., B.S., Koç, Ç.K., Paar, C. (eds.) CHES 2002. LNCS, vol. 2523, pp. 291–302. Springer, Heidelberg (2003)
14. Kim, C.H., Quisquater, J.: How can we overcome both side channel analysis and fault attacks on RSA-CRT? In: Fault Diagnosis and Tolerance in Cryptography (FDTC 2007), pp. 21–29. IEEE (2007)
15. Kiss, A., Krämer, J., Rauzy, P., Seifert, J.P.: Algorithmic countermeasures against fault attacks and power analysis for RSA-CRT. Cryptology ePrint Archive, Report 2016/238 (2016). http://eprint.iacr.org/2016/238
16. Kocher, P.C., Jaffe, J., Jun, B.: Differential power analysis. In: Wiener, M. (ed.) CRYPTO 1999. LNCS, vol. 1666, pp. 388–397. Springer, Heidelberg (1999)
17. Krämer, J., Nedospasov, D., Seifert, J.-P.: Weaknesses in current RSA signature schemes. In: Kim, H. (ed.) ICISC 2011. LNCS, vol. 7259, pp. 155–168. Springer, Heidelberg (2012)
18. Le, D.-P., Rivain, M., Tan, C.H.: On double exponentiation for securing RSA against fault analysis. In: Benaloh, J. (ed.) CT-RSA 2014. LNCS, vol. 8366, pp. 152–168. Springer, Heidelberg (2014)

19. Montgomery, P.L.: Speeding the Pollard and elliptic curve methods of factorization. Math. Comput. **48**(177), 243–264 (1987)
20. Quisquater, J.J., Couvreur, C.: Fast decipherment algorithm for RSA public-key cryptosystem. Electron. Lett. **18**(21), 905–907 (1982)
21. Rauzy, P., Guilley, S.: Countermeasures against high-order fault-injection attacks on CRT-RSA. In: Fault Diagnosis and Tolerance in Cryptography (FDTC 2014), pp. 68–82. IEEE (2014)
22. Rivain, M.: Securing RSA against fault analysis by double addition chain exponentiation. In: Fischlin, M. (ed.) CT-RSA 2009. LNCS, vol. 5473, pp. 459–480. Springer, Heidelberg (2009)
23. Rivest, R.L., Shamir, A., Adleman, L.M.: A method for obtaining digital signatures and public-key cryptosystems. Commun. ACM **21**(2), 120–126 (1978)
24. Shamir, A.: Method and apparatus for protecting public key schemes from timing and fault attacks, US Patent 5,991,415 (1999)
25. Witteman, M.: A DPA attack on RSA in CRT mode (2009)
26. Yen, S., Joye, M.: Checking before output may not be enough against fault-based cryptanalysis. IEEE Trans. Comput. **49**(9), 967–970 (2000)
27. Yen, S.-M., Kim, S., Lim, S., Moon, S.-J.: A countermeasure against one physical cryptanalysis may benefit another attack. In: Kim, K. (ed.) ICISC 2001. LNCS, vol. 2288, pp. 414–427. Springer, Heidelberg (2002)
28. Yen, S.-M., Lien, W.-C., Moon, S.-J., Ha, J.C.: Power analysis by exploiting chosen message and internal collisions – vulnerability of checking mechanism for RSA-decryption. In: Dawson, E., Vaudenay, S. (eds.) Mycrypt 2005. LNCS, vol. 3715, pp. 183–195. Springer, Heidelberg (2005)

Improved Differential Fault Analysis on Camellia-128

Toru Akishita[✉] and Noboru Kunihiro

The University of Tokyo, Tokyo, Japan
Toru.Akishita@jp.sony.com, kunihiro@k.u-tokyo.ac.jp

Abstract. In this paper we propose improved Differential Fault Analysis (DFA) on the block cipher Camellia with a 128-bit key. Existing DFAs on Camellia-128 require several faults induced at multiple rounds, at least two of which must be induced at or after the 16-th round. On the other hand, by utilizing longer fault propagation paths than the existing DFAs, the proposed attacks require random byte faults to targeted byte positions induced only at the 14-th round. The simulation results confirm the feasibility of the proposed attacks. Our attacks indicate that the last 5 rounds of Camellia-128, two more rounds compared with the existing DFAs, must be protected against DFAs.

Keywords: Differential fault analysis · DFA · Camellia · Fault propagation path

1 Introduction

The idea of fault analysis was first proposed by Boneh et al. [5]. They showed that the secret key of RSA-CRT implementations can be easily detected by a pair of a correct and a faulty plaintext for an identical ciphertext. Shortly after, Biham and Shamir [4] proposed Differential Fault Analysis (DFA) against symmetric key cryptography which combines the concept of differential analysis and fault analysis. They showed that the entire key of DES can be retrieved by between 50 and 200 faulty ciphertexts. Since then various DFA attacks have been proposed to block ciphers such as AES [8], Camellia [3], and CLEFIA [13].

Various DFAs on block ciphers have been proposed aiming at mainly the following two directions: minimizing the number of faults and making the location of fault injections earlier round of each cipher. With respect to the former direction, the best DFAs has been proposed using a single fault for AES [1,16] and two faults for CLEFIA [14]. Regarding the latter direction, there is a trade-off between protection and efficiency. Designers are motivated to limit the number of rounds to protect in order to save computational cost. Therefore, it is important to investigate how many rounds of each cipher we need to protect against DFA. Several DFAs exploiting faults at an early round have been proposed for AES [9,11,12] and CLEFIA [2,15]. This paper presents the first improvement in the latter direction on Camellia.

© Springer International Publishing Switzerland 2016
F.-X. Standaert and E. Oswald (Eds.): COSADE 2016, LNCS 9689, pp. 130–143, 2016.
DOI: 10.1007/978-3-319-43283-0_8

Camellia is a 128-bit block cipher jointly developed by Mitsubishi Electric Corporation and NTT in 2000 [3]. Camellia was selected as a recommended cryptographic primitive by the EU NESSIE project [10] and was included in Japanese "e-Government Recommended Ciphers List" selected by Japan CRYPTREC [7]. Also, it is one of the three 128-bit block ciphers in ISO/IEC 18033-3. For Camellia, several DFAs have been proposed. We summarize the results of the proposed attacks on Camellia with a 128-bit key in Table 1. In [17], Zhao et al. proposed the DFA which repeatedly induces random byte faults from the 17-th round to the 14-th round. Using 16 pairs of correct and faulty ciphertexts the attack retrieves the 128-bit secret key. The attack is improved independently by Chen et al. [6] and Zhao et al. [18]. Both attacks require random byte faults at the 16-th round and the 14-th round, while the number of faults is different; Chen's attack requires 6 faulty ciphertexts and Zhao's attack requires 4 faulty ciphertexts. To the best of our knowledge, all the existing DFAs on Camellia with a 128-bit key require several faults induced at multiple rounds, at least two of which must be induced at or after the 16-th round. Therefore, the last 3 rounds require to be protected against existing DFAs.

Table 1. Comparison of DFAs against Camellia-128

Reference	Fault location	# of faults	Complexity
[17]	14/15/16/17-th round	16	-
[6]	14/16-th round	6	-
[18]	14/16-th round	4	$2^{22.2}$
Ours	14-th round	8	$2^{9.20}$
		7	$2^{16.14}$
		6	$2^{23.12}$
		5	$2^{30.10}$
		4	$2^{37.08}$

In this paper, we propose improved DFAs on Camellia with a 128-bit key, inspired by DFAs on CLEFIA proposed by Todo and Sasaki [15]. We utilize longer fault propagation paths than the existing DFAs. Table 1 shows comparison of the existing DFAs and our DFAs. Our DFAs can exploit random byte faults induced only at the 14-th round, which is two round earlier than that of existing DFAs. On the other hand, we mount our attack under an additional condition of fault model, where an attacker can induce random byte faults to the targeted byte position. The complexity of our attacks depends on the number of faults; our DFA with 8 faults require the complexity of $2^{9.20}$, while our DFA with 4 faults require the complexity of $2^{37.08}$. We performed a simulation on a PC to estimate the calculation time of our DFA with 8 faults. The average time for retrieving the key of 10,000 samples is 0.945 ms.

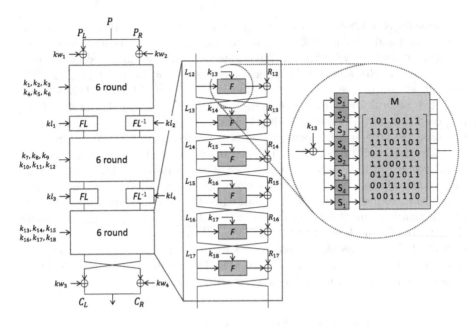

Fig. 1. Encryption structure of Camellia-128

The rest of the paper is organized as follows. Sect. 2 gives brief description of Camellia and previous DFAs on Camellia. We show the proposed attack with 8 faults in Sect. 3 and its simulation results in Sect. 4. We reduce the number of faults down to 4 in Sect. 5. Finally, we conclude in Sect. 6.

2   Preliminaries

2.1   Block Cipher Camellia

Camellia [3] is a 128-bit block cipher with its key length 128, 192, and 256 bits. In the paper, we consider 128-bit key Camellia, denoted as Camellia-128, though similar techniques are applicable to Camellia with 192-bit and 256-bit keys. Camellia-128 uses an 18-round Feistel structure with key whitenings and logical functions called the FL-function and FL^{-1}-function inserted every 6 rounds. Fig. 1 shows encryption structure of Camellia-128. Let $P, C \in \{0,1\}^{128}$ be a plaintext and a ciphertext, and we divide P and C into $P = P_L|P_R$ and $C = C_L|C_R$, where $P_L, P_R, C_L, C_R \in \{0,1\}^{64}$. Let $L_{i-1}, R_{i-1} \in \{0,1\}^{64}$ be the left half and the right half of the i-th round input data, and $k_i \in \{0,1\}^{64}$ be the subkey at i-th round. Let $kw_t \in \{0,1\}^{64}$ for $t = 1, \ldots, 4$ be the subkey for Prewhitening or Postwhitening, and $kl_v \in \{0,1\}^{64}$ for $v = 1, \ldots, 4$ be the subkey for FL/FL^{-1}-function.

The F-function F consists of subkey addition, 8 non-linear 8-bit S-boxes and a diffusion matrix. The construction of F is shown in Fig. 1. Four kind of S-boxes S_1, S_2, S_3 and S_4, where $S_1, S_2, S_3, S_4 : \{0,1\}^8 \to \{0,1\}^8$, are employed.

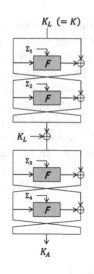

	subkey	value
F (Round 15)	k_{15}	$(K_A \lll_{94})_L$
F (Round 16)	k_{16}	$(K_A \lll_{94})_R$
F (Round 17)	k_{17}	$(K_L \lll_{111})_L$
F (Round 18)	k_{18}	$(K_L \lll_{111})_R$
Postwhitening	kw_3	$(K_A \lll_{111})_L$
	kw_4	$(K_A \lll_{111})_R$

Fig. 2. Key scheduling of Camellia-128

The diffusion matrix M is a binary matrix and updates eight 8-bit values $(x_1, x_2, x_3, x_4, x_5, x_6, x_7, x_8)$ as follows:

$$^t(x_1, x_2, x_3, x_4, x_5, x_6, x_7, x_8) \leftarrow M \cdot {}^t(x_1, x_2, x_3, x_4, x_5, x_6, x_7, x_8),$$

where the operation between matrices and vectors performed over GF(2).

All the subkeys are generated by the key scheduling algorithm. The key scheduling algorithm of Camellia-128 generates an 128-bit variable K_A from K_L that is equal to the 128-bit secret key K as shown in Fig. 2, where Σ_i for $i = 1, \ldots, 4$ are constant values. We also list the value of subkeys for the last 4 rounds and postwhitening in Fig. 2. For Camellia-128, attackers can derive the 128-bit secret key K from $k'_{15}, k'_{16}, k'_{17}, k'_{18}$, where $k'_{15}, k'_{16}, k'_{17}$ and k'_{18} denote $k_{15} \oplus kw_4, k_{16} \oplus kw_3, k_{17} \oplus kw_4$ and $k_{18} \oplus kw_3$, respectively.

2.2 Previous DFAs on Camellia-128

In this section we briefly explain the existing two DFAs on Camellia-128. Both DFAs employed a method that narrows down the candidates for 8 bits of an unknown subkey by solving the equation for an S-box as follows.

Let us consider the basic one-byte S-box model, where one-byte input x are XOR-ed to one-byte key k, and then the result are input to S-box $S[]$ and output $S[x \oplus k]$. When we know a pair of inputs x_a and x_b, and know output differential δ, we can obtain a set of unknown key candidates by solving the following equation:

$$S[x_a \oplus k] \oplus S[x_b \oplus k] \oplus \delta = 0.$$

The size of key candidate spaces depends on x_a, x_b, and δ, and the structure of the S-box. In the case of the S-boxes S_1, S_2, S_3 and S_4 of Camellia, the same as $S1$ of CLEFIA [14], expected value of the size of keys candidate spaces is $2^{1.02}$.

In [17], attackers induce random byte faults at L_{16}. Since a byte fault enables to deduce key candidates for 5 or 6 bytes of k'_{18}, the authors argue that 4 random faults are enough to recover k'_{18} with a high probability. In the same way attackers can recover k'_{17}, k'_{16} and k'_{15} by inducing 4 random byte faults at L_{15}, L_{14} and L_{13} accordingly, and finally they get the secret key K with 16 faults.

The improved attack proposed in [18] successfully reduce the number of faults by extending fault depth. Attackers induce a random byte fault at L_{15} to recover 8 bytes of k'_{18}, 5 or 6 bytes of k'_{17} and 1 byte of k'_{16}. The authors argue that 2 faults are enough to recover k'_{18} and k'_{17} with high probability. In the same way attackers can recover k'_{16} and k'_{15} by inducing 2 random byte faults at L_{13}, and finally they get the secret key K with 4 faults. The complexity of the attack is estimated to be $2^{22.2}$.

3 Proposed DFA on Camellia-128

In this section, we propose a new DFA on Camellia-128 by utilizing 5-round fault propagation paths. Here we show our original attack inducing 8 random byte faults at R_{13}, The attack consists of 6 steps. Note that $Z[i]$ denotes the i-th byte of 8-byte Z.

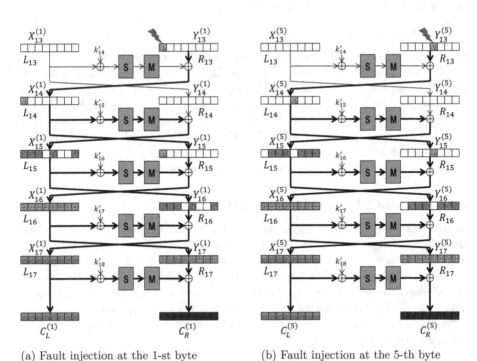

(a) Fault injection at the 1-st byte   (b) Fault injection at the 5-th byte

Fig. 3. Fault propagation path in the last 5 rounds

Step 1: Inducing Faults at R_{13}. In our DFA, attackers induce 8 random byte faults as follows. Attackers first get a correct ciphertext C corresponding to a plaintext. Next attackers induce random byte fault to the 1-st byte of R_{13} in the encryption of the same plaintext and secret key, and get a faulty ciphertext $C^{(1)} = C_L^{(1)}|C_R^{(1)}$. The fault at the 1-st byte of R_{13} is propagated in the last 5 rounds as shown in Fig. 3(a). We move post-whitening keys wk_3 and wk_4 to the positions of round keys k_i by an equivalent transformation, where $k'_{14}, k'_{15}, k'_{16}, k'_{17}$ and k'_{18} denote $k_{14} \oplus kw_3, k_{15} \oplus kw_4, k_{16} \oplus kw_3, k_{17} \oplus kw_4$ and $k_{18} \oplus kw_3$, respectively. X_i and Y_i denote the values of L_i and R_i respectively after the equivalent transformation.

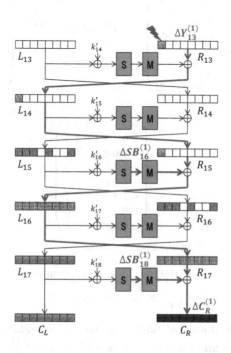

Fig. 4. 5-round paths for the 1-st byte fault of R_{13}

Similarly, attackers induce random byte faults to the i-th byte of R_{13}, and get faulty ciphertexts $C^{(i)} = C_L^{(i)}|C_R^{(i)}$ for $i = 2, \ldots, 8$. Fig. 3(b) show the fault propagation paths when inducing it at the 5-th byte of R_{13}.

Let $SB_j \in \{0,1\}^{64}$ be the output of S-box layer at the j-th round. $X_j^{(i)}, Y_j^{(i)}$ and $SB_j^{(i)}$ denote X_j, Y_j and SB_j, respectively, when inducing faults at the i-th byte of R_{13}.

Step 2: Narrowing Down Candidates of k'_{18}. In this step, attackers narrow down candidates of k'_{18} to $2^{4.08}$ from 2^{64}. Firstly we show how to reduce the key candidates of k'_{18} by using the correct ciphertext C and the faulty ciphertexts $C^{(1)}$. By utilizing the 5-round paths as shown in Fig. 4, we have the following differential equation:

$$\Delta Y_{13}^{(1)} \oplus M(\Delta SB_{16}^{(1)}) \oplus M(\Delta SB_{18}^{(1)}) = \Delta C_R^{(1)},$$

where $\Delta Y_{13}^{(1)} = Y_{13}^{(1)} \oplus Y_{13}$, $\Delta SB_{16}^{(1)} = SB_{16}^{(1)} \oplus SB_{16}$, $\Delta SB_{18}^{(1)} = SB_{18}^{(1)} \oplus SB_{18}$, and $\Delta C_R^{(1)} = C_R^{(1)} \oplus C_R$. Since each byte of $\Delta Y_{13}^{(1)}$ is equal to 0 except the 1-st byte and the 4-th, 6-th and 7-th byte of $\Delta SB_{16}^{(1)}$ are equal to 0, it satisfies that

$$
\begin{pmatrix}
\Delta SB_{16}^{(1)}[1] \oplus \Delta SB_{18}^{(1)}[1] \\
\Delta SB_{16}^{(1)}[2] \oplus \Delta SB_{18}^{(1)}[2] \\
\Delta SB_{16}^{(1)}[3] \oplus \Delta SB_{18}^{(1)}[3] \\
\Delta SB_{18}^{(1)}[4] \\
\Delta SB_{16}^{(1)}[5] \oplus \Delta SB_{18}^{(1)}[5] \\
\Delta SB_{18}^{(1)}[6] \\
\Delta SB_{18}^{(1)}[7] \\
\Delta SB_{16}^{(1)}[8] \oplus \Delta SB_{18}^{(1)}[8]
\end{pmatrix}
= M^{-1}\Delta C_R^{(1)} \oplus M^{-1}
\begin{pmatrix}
\Delta Y_{13}^{(1)} \\
0 \\
0 \\
0 \\
0 \\
0 \\
0 \\
0
\end{pmatrix},
$$

where

$$
M^{-1} =
\begin{pmatrix}
0 1 1 1 0 1 1 1 \\
1 0 1 1 1 0 1 1 \\
1 1 0 1 1 1 0 1 \\
1 1 1 0 1 1 1 0 \\
1 1 0 0 1 0 1 1 \\
0 1 1 0 1 1 0 1 \\
0 0 1 1 1 1 1 0 \\
1 0 0 1 0 1 1 1
\end{pmatrix}.
$$

Therefore, we have the following two equations:

$$\Delta SB_{18}^{(1)}[6] = (M^{-1}\Delta C_R^{(1)})[6],$$
$$\Delta SB_{18}^{(1)}[7] = (M^{-1}\Delta C_R^{(1)})[7].$$

Attackers can narrow down candidates of the corresponding key $k_{18}'[6]$ to $2^{1.02}$ from 2^8 because they know a pair of inputs $C_L[6]$ and $C_L^{(1)}[6]$, and output differential $\Delta SB_{18}^{(1)}[6]$ as described in Sect. 2.2. Attackers can also reduce the key space of $k_{18}'[7]$ to $2^{1.02}$.

Similarly attackers can narrow down candidates of $k_{18}'[7]$ and $k_{18}'[8]$ by using the correct ciphertext C and the faulty ciphertext $C^{(2)}$. We summarize the relationship between fault positions of R_{13} and their corresponding keys $k_{18}'[l]$ whose key space is reduced to $2^{1.02}$ in Table 2.

Since the candidates of $k_{18}'[5]$, $k_{18}'[6]$, $k_{18}'[7]$ and $k_{18}'[8]$ are narrowed down twice respectively shown in Table 2, these keys are uniquely determined with a high probability. Therefore, we can reduce key space of k_{18}' to $(2^{1.02})^4 = 2^{4.08}$ from 2^{64} in this step.

Step 3: Recovering k'_{18} and k'_{17}. In this step, attackers uniquely determine k_{18}' and k_{17}'. Firstly we show how to reduce the key candidates of k_{18}' and k_{17}' by using the correct ciphertext C and the faulty ciphertexts $C^{(1)}$. Attackers compute

Table 2. Fault position of R_{13} and deduced keys at step 2

Fault position of R_{13}	Keys
1-st byte	$k'_{18}[6]$, $k'_{18}[7]$
2-nd byte	$k'_{18}[7]$, $k'_{18}[8]$
3-rd byte	$k'_{18}[5]$, $k'_{18}[8]$
4-th byte	$k'_{18}[5]$, $k'_{18}[6]$
5-th byte	$k'_{18}[1]$
6-th byte	$k'_{18}[2]$
7-th byte	$k'_{18}[3]$
8-th byte	$k'_{18}[4]$

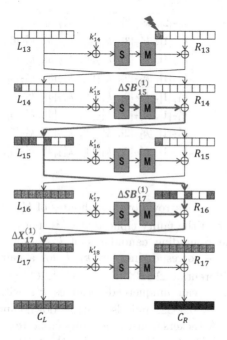

Fig. 5. 3-round paths for the 1-st byte fault of R_{13}

$2^{4.08}$ set of $\{Y_{17}, Y_{17}^{(1)}\}$ for $2^{4.08}$ candidates of k'_{18}. By utilizing the 3-round paths as shown in Fig. 5, we have the following differential equation:

$$M(\Delta SB_{15}^{(1)}) \oplus M(\Delta SB_{17}^{(1)}) = \Delta X_{17}^{(1)},$$

where $\Delta X_{17}^{(1)}$ is identical to $\Delta C_L^{(1)}$. Since each byte of $\Delta SB_{15}^{(1)}$ is equal to 0 except the 1-st byte, we have the following seven equations:

$$\Delta SB_{17}^{(1)}[l] = (M^{-1}\Delta C_L^{(1)})[l],$$

for $l = 2, \ldots, 8$. For each set of $\{Y_{17}, Y_{17}^{(1)}\}$, attacker confirm that there exists candidates of the corresponding key $k'_{17}[l]$ for a pair of inputs $Y_{17}[l]$ and $Y_{17}^{(1)}[l]$, and output differential $\Delta SB_{17}^{(1)}[l]$, where $2 \leq l \leq 8$. For an incorrect candidate of $(Y_{17}, Y_{17}^{(1)})$, the probability that all $k'_{17}[l]$ exist for $l = 2, \ldots, 8$ is $(2^{-1.02})^7 = 2^{-7.14}$. Thus the probability that all incorrect key candidates of k'_{18} are eliminated, namely, k'_{18} is uniquely determined is equal to $(1 - 2^{-7.14})^{(2^{4.08}-1)} = 0.893$. Attackers can also narrow down candidates of $k'_{17}[l]$ for $l = 2, \ldots, 8$ to $2^{1.02}$ from 2^8.

Table 3. Fault position of R_{13} and deduced keys at Step 3

Fault position of R_{13}	Keys
1-st byte	$k'_{17}[2], \ k'_{17}[3], \ k'_{17}[4], \ k'_{17}[5], \ k'_{17}[6], \ k'_{17}[7], \ k'_{17}[8]$
2-nd byte	$k'_{17}[1], \ k'_{17}[3], \ k'_{17}[4], \ k'_{17}[5], \ k'_{17}[6], \ k'_{17}[7], \ k'_{17}[8]$
3-rd byte	$k'_{17}[1], \ k'_{17}[2], \ k'_{17}[4], \ k'_{17}[5], \ k'_{17}[6], \ k'_{17}[7], \ k'_{17}[8]$
4-th byte	$k'_{17}[1], \ k'_{17}[2], \ k'_{17}[3], \ k'_{17}[5], \ k'_{17}[6], \ k'_{17}[7], \ k'_{17}[8]$
5-th byte	$k'_{17}[1], \ k'_{17}[2], \ k'_{17}[3], \ k'_{17}[4], \ k'_{17}[6], \ k'_{17}[7], \ k'_{17}[8]$
6-th byte	$k'_{17}[1], \ k'_{17}[2], \ k'_{17}[3], \ k'_{17}[4], \ k'_{17}[5], \ k'_{17}[7], \ k'_{17}[8]$
7-th byte	$k'_{17}[1], \ k'_{17}[2], \ k'_{17}[3], \ k'_{17}[4], \ k'_{17}[5], \ k'_{17}[6], \ k'_{17}[8]$
8-th byte	$k'_{17}[1], \ k'_{17}[2], \ k'_{17}[3], \ k'_{17}[4], \ k'_{17}[5], \ k'_{17}[6], \ k'_{17}[7]$

Similarly attackers can narrow down candidates of k'_{18} and k'_{17} by using the correct ciphertext C and the faulty ciphertext $C^{(2)}$. For each set of $\{Y_{17}, Y_{17}^{(2)}\}$ corresponding to the reminding candidates of k'_{18}, they confirm that there exists candidates of the corresponding key $k'_{17}[l]$ for a pair of inputs Y_{17} and $Y_{17}^{(2)}$, and output differential $\Delta SB_{17}^{(2)}[l] = (M^{-1}\Delta C_L^{(2)})[l]$, where $l = 1$ or $3 \leq l \leq 8$. Attackers can uniquely determines k'_{18} with high probability, $(1 - (2^{-7.14})^2)^{(2^{4.08}-1)} = 0.9992$, and also narrow down candidates of $k'_{17}[l]$ for $l = 1$ and $l = 3, \ldots, 8$. Attackers utilizes the faulty ciphertext $C^{(i)}$ for $i = 3, \ldots, 8$ until uniquely determining k'_{17}. We summarize the relationship between fault positions of R_{13} and their corresponding keys $k'_{17}[l]$ whose key space is reduced to $2^{1.02}$ in Table 3.

Step 4: Recovering k'_{16}. In this step, attackers can uniquely determine k'_{16}. Firstly attackers compute $\{X_{16}, Y_{16}\}$ and $\{X_{16}^{(1)}, Y_{16}^{(1)}\}$ using k'_{18} and k'_{17}. We have following differential equation:

$$\Delta Y_{13}^{(1)} \oplus M(\Delta SB_{16}^{(1)}) = \Delta X_{16}^{(1)},$$

Since the 4-th, 6-th and 7-th byte of $\varDelta SB_{16}^{(1)}$ are equal to 0, we have the following six equations:

$$\varDelta SB_{16}^{(1)}[1] = (M^{-1}\varDelta X_{16}^{(1)})[1],$$
$$\varDelta SB_{16}^{(1)}[2] = (M^{-1}\varDelta X_{16}^{(1)})[2] \oplus \varDelta Y_{13}^{(1)},$$
$$\varDelta SB_{16}^{(1)}[3] = (M^{-1}\varDelta X_{16}^{(1)})[3] \oplus \varDelta Y_{13}^{(1)},$$
$$0 = (M^{-1}\varDelta X_{16}^{(1)})[4] \oplus \varDelta Y_{13}^{(1)},$$
$$\varDelta SB_{16}^{(1)}[5] = (M^{-1}\varDelta X_{16}^{(1)})[5] \oplus \varDelta Y_{13}^{(1)},$$
$$\varDelta SB_{16}^{(1)}[8] = (M^{-1}\varDelta X_{16}^{(1)})[8] \oplus \varDelta Y_{13}^{(1)}.$$

Thus attackers can narrow down candidates of $k_{16}'[l]$ for $l = 1, 2, 3, 5, 8$ to $2^{1.02}$ from 2^8. Attackers utilizes the faulty ciphertext $C^{(i)}$ for $i = 2, \ldots, 8$ until uniquely determining k_{16}'. We summarize the relationship between fault positions of R_{13} and their corresponding keys $k_{16}'[l]$ whose key space is reduced to $2^{1.02}$ in Table 4.

Table 4. Fault position of R_{13} and deduced keys at Step 4

Fault position of R_{13}	Keys
1-st byte	$k_{16}'[1]$, $k_{16}'[2]$, $k_{16}'[3]$, $k_{16}'[5]$, $k_{16}'[8]$
2-nd byte	$k_{16}'[2]$, $k_{16}'[3]$, $k_{16}'[4]$, $k_{16}'[5]$, $k_{16}'[6]$
3-rd byte	$k_{16}'[1]$, $k_{16}'[3]$, $k_{16}'[4]$, $k_{16}'[6]$, $k_{16}'[7]$
4-th byte	$k_{16}'[1]$, $k_{16}'[2]$, $k_{16}'[4]$, $k_{16}'[7]$, $k_{16}'[8]$
5-th byte	$k_{16}'[2]$, $k_{16}'[3]$, $k_{16}'[4]$, $k_{16}'[6]$, $k_{16}'[7]$, $k_{16}'[8]$
6-th byte	$k_{16}'[1]$, $k_{16}'[3]$, $k_{16}'[4]$, $k_{16}'[5]$, $k_{16}'[7]$, $k_{16}'[8]$
7-th byte	$k_{16}'[1]$, $k_{16}'[2]$, $k_{16}'[4]$, $k_{16}'[5]$, $k_{16}'[6]$, $k_{16}'[8]$
8-th byte	$k_{16}'[1]$, $k_{16}'[2]$, $k_{16}'[3]$, $k_{16}'[5]$, $k_{16}'[6]$, $k_{16}'[7]$

Step 5: Narrowing Down Candidates of k'_{15}. In this step, attackers narrow down candidates of k_{15}' to $2^{8.16}$ from 2^{64}. Firstly attacker computes Y_{15} and $Y_{15}^{(1)}$. We have following differential equation:

$$M(\varDelta SB_{15}^{(1)}) = \varDelta X_{15}^{(1)},$$

where $\varDelta X_{15}^{(1)}$ is identical to $\varDelta Y_{16}^{(1)}$. Since each byte of $\varDelta SB_{15}^{(1)}$ is equal to 0 except the 1-st byte, we have the following equation:

$$\varDelta SB_{15}^{(1)}[1] = (M^{-1}\varDelta Y_{16}^{(1)})[1].$$

Thus attackers can narrow down candidates of $k_{15}'[1]$ to $2^{1.02}$ from 2^8. Similarly attackers can narrow down candidates $k_{15}'[i]$ to $2^{1.02}$ from 2^8 by utilizing the faulty ciphertexts $C^{(i)}$ respectively for $i = 2, \ldots, 8$. We can reduce key space of k'_{15} to $(2^{1.02})^8 = 2^{8.16}$ from 2^{64} in this step.

Step 6: Recovering the Secret Key K. From step 1 to step 5, attackers recover k'_{18}, k'_{17} and k'_{16}, and get $2^{8.16}$ candidates of k'_{15}. We have $2^{8.16} \times 2 = 2^{9.16}$ candidates of $\{K_A, K_L\}$ because the key scheduling of Camellia-128 satisfy the following equation:

$$k'_{15}|k'_{16} = (k_{15} \oplus kw_4)|(k_{16} \oplus kw_3) = (K_A \lll_{94}) \oplus (K_A \lll_{47}),$$
$$k'_{17}|k'_{18} = (k_{17} \oplus kw_4)|(k_{18} \oplus kw_3) = (K_L \lll_{111}) \oplus (K_A \lll_{47}).$$

For each set of $\{K_A, K_L\}$, attackers generate an 128-bit variable K'_A from K_L as shown in Fig. 2 and check whether K'_A is equal to K_A. If K'_A is equal to K_A, the corresponding K_L is the secret key K.

Complexity Analysis. The complexity of step 3 and step 6 is $2^{4.08}$ and $2^{9.16}$, respectively. The complexity of the other steps is negligible. Thus the complexity of the proposed attack is about $2^{9.20}$.

4   Simulation Results

In order to verify the feasibility of the proposed attack, we implemented the simulation of our attack in C code and executed on a single core in an Intel Core i5-3210M 2.5 GHz notebook machine. In the simulation, we used random 128-bit keys and plaintexts, and then induced 8 random byte faults at R_{13} during encryption. The histogram of calculation time of 10,000 samples is shown in Fig. 6. We successfully retrieved the 128-bit secret key in all of the 10,000 samples. The average time and the longest one is 0.945 ms and 116 ms, respectively. The total calculation time for retrieving the key is less than 1.0 ms for 81.6 % and less than 4.0 ms for 99.9 %.

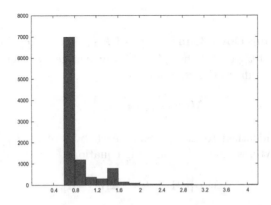

Fig. 6. Histogram of the calculation time of 10,000 samples

5   Reducing the Number of Faults

In this section, we try to reduce the number of faults for our original DFA proposed in Sect. 3.

Let us consider that we omit a fault injection at the 1-st byte of R_{13}. In step 2, key space of k'_{18} is $(2^{1.02})^6 = 2^{6.12}$ because attackers cannot uniquely determine $k'_{18}[6]$ and $k'_{18}[7]$. In step 3, attackers can uniquely determine k'_{18} and k'_{17}, and its complexity is $2^{6.12}$. In step 5, attackers cannot narrow down candidates of $k'_{15}[1]$ from 2^8, and then key space of k'_{15} is $2^8 \times (2^{1.02})^7 = 2^{15.14}$. The complexity of step 6 is $2^{15.14} \times 2 = 2^{16.14}$. Thus, the complexity of the attack with 7 faults is about $2^{16.14}$.

Let us consider that we induce a random byte fault at the 1-st, 3-rd, 5-th and 6-th byte of R_{13}. In step 2, attackers can narrow down candidates of $k'_{18}[1]$, $k'_{18}[2]$, $k'_{18}[5]$, $k'_{18}[6]$, $k'_{18}[7]$ and $k'_{18}[8]$ to $2^{1.02}$, but cannot narrow candidates of $k'_{18}[3]$ and $k'_{18}[4]$ from 2^8. Key space of k'_{18} is $(2^{1.02})^6 \times (2^8)^2 = 2^{22.12}$. In step 3, attackers can uniquely determine k'_{18} with the probability of $(1 - (2^{-7.14})^4)^{(2^{22.12}-1)} = 0.9885$. Thus, attacker can uniquely determine k'_{18} and k'_{17} with a high probability, and its complexity is $2^{22.12}$. In step 5, attackers cannot narrow down candidates of $k'_{15}[2]$, $k'_{15}[4]$, $k'_{15}[7]$ and $k'_{15}[8]$ from 2^8, and then key space of k'_{15} is $(2^8)^4 \times (2^{1.02})^4 = 2^{36.08}$. The complexity of step 6 is $2^{36.08} \times 2 = 2^{37.08}$. Thus, the complexity of the attack with 4 faults is about $2^{37.08}$.

We summarize our optimized DFAs with the number of faults being from 4 to 8 and their complexity in Table 5.

Table 5. Our optimized DFAs and complexity

# of faults	Fault position of R_{13}	Complexity		
		Step 3	Step 6	Total
8	1-st, 2-nd, 3-rd, 4-th	$2^{4.08}$	$2^{9.16}$	$2^{9.20}$
	5-th, 6-th, 7-th, 8-th			
7	three of {1-st, 2-nd, 3-rd, 4-th}	$2^{6.12}$	$2^{16.14}$	$2^{16.14}$
	5-th, 6-th, 7-th, 8-th			
6	{1-st, 3-rd} or {2-rd, 4-th}	$2^{8.16}$	$2^{23.12}$	$2^{23.12}$
	5-th, 6-th, 7-th, 8-th			
5	{1-st, 3-rd} or {2-rd, 4-th}	$2^{15.14}$	$2^{30.10}$	$2^{30.10}$
	three of {5-th, 6-th, 7-th, 8-th}			
4	{1-st, 3-rd} or {2-rd, 4-th}	$2^{22.12}$	$2^{37.08}$	$2^{37.08}$
	two of {5-th, 6-th, 7-th, 8-th}			

6   Conclusion

In the paper we propose improved DFAs on Camellia-128. Our attacks can exploit random byte faults induced only at the 14-th round, which is two round earlier than that of existing DFAs. The complexity of our attacks depends on the number of faults; our DFA with 8 faults require the complexity of $2^{9.20}$, while our DFA with 4 faults require the complexity of $2^{37.08}$. Our simulation results confirm the feasibility of the proposed attack with 8 faults. We argue that the last 5 rounds, not the last 3 rounds, of Camellia must be protected against DFAs.

Acknowledgment. We would like to thank the anonymous reviewers for their helpful comments. This research was partially supported by CREST, JST and JSPS KAK-ENHI Grant Number 25280001.

References

1. Ali, S.S., Mukhopadhyay, D.: A differential fault analysis on aes key schedule using single fault. In: FDTC, pp. 35–42. IEEE (2011)
2. Ali, S.S., Mukhopadhyay, D.: Improved differential fault analysis of CLEFIA. In: The 10th Workshop on Fault Diagnosis and Tolerance in Cryptography - FDTC, pp. 60–70. IEEE (2013)
3. Aoki, K., Ichikawa, T., Kanda, M., Matsui, M., Moriai, S., Nakajima, J., Tokita, T.: *Camellia*: a 128-bit block cipher suitable for multiple platforms - design and analysis. In: Stinson, D.R., Tavares, S. (eds.) SAC 2000. LNCS, vol. 2012, p. 39. Springer, Heidelberg (2001)
4. Biham, E., Shamir, A.: Differential fault analysis of secret key cryptosystems. In: Kaliski Jr., B.S. (ed.) CRYPTO 1997. LNCS, vol. 1294, pp. 513–525. Springer, Heidelberg (1997)
5. Boneh, D., DeMillo, R.A., Lipton, R.J.: On the importance of checking cryptographic protocols for faults. In: Fumy, W. (ed.) EUROCRYPT 1997. LNCS, vol. 1233, pp. 37–51. Springer, Heidelberg (1997)
6. Chen, H., Zhou, Y., Wu, W., Wang, N.: Fault propagation pattern based DFA on feistel ciphers, with application to Camellia. In: The 10th IEEE International Conference on Computer and Information Technology - CIT, pp. 1050–1057. IEEE Computer Society (2010)
7. Japan CRYPTREC (Cryptography Research and Evaluation Committees). http://www.cryptrec.go.jp/english/index.html
8. Daemen, J., Rijmen, V.: The Design of Rijndael: AES - The Advanced Encryption Standard. Information Security and Cryptography, 1st edn. Springer, Heidelberg (2002)
9. Derbez, P., Fouque, P.-A., Leresteux, D.: Meet-in-the-middle and impossible differential fault analysis on AES. In: Preneel, B., Takagi, T. (eds.) CHES 2011. LNCS, vol. 6917, pp. 274–291. Springer, Heidelberg (2011)
10. The NESSIE project (New European Schemes for Signatures, Integrity and Encryption). https://www.cosic.esat.kuleuven.be/nessie/
11. Phan, R.C.-W., Yen, S.-M.: Amplifying side-channel attacks with techniques from block cipher cryptanalysis. In: Domingo-Ferrer, J., Posegga, J., Schreckling, D. (eds.) CARDIS 2006. LNCS, vol. 3928, pp. 135–150. Springer, Heidelberg (2006)
12. Sasaki, Y., Li, Y., Sakamoto, H., Sakiyama, K.: Coupon collector's problem for fault analysis against AES — high tolerance for noisy fault injections. In: Sadeghi, A.-R. (ed.) FC 2013. LNCS, vol. 7859, pp. 213–220. Springer, Heidelberg (2013)

13. Shirai, T., Shibutani, K., Akishita, T., Moriai, S., Iwata, T.: The 128-bit blockcipher CLEFIA (Extended Abstract). In: Biryukov, A. (ed.) FSE 2007. LNCS, vol. 4593, pp. 181–195. Springer, Heidelberg (2007)
14. Takahashi, J., Fukunaga, T.: Improved differential fault analysis on CLEFIA. In: The 5th Workshop on Fault Diagnosis and Tolerance in Cryptography - FDTC, pp. 25–34. IEEE (2008)
15. Todo, Y., Sasaki, Y.: New property of diffusion switching mechanism on CLEFIA and its application to DFA. In: Sakiyama, K., Terada, M. (eds.) IWSEC 2013. LNCS, vol. 8231, pp. 99–114. Springer, Heidelberg (2013)
16. Tunstall, M., Mukhopadhyay, D., Ali, S.: Differential fault analysis of the advanced encryption standard using a single fault. In: Ardagna, C.A., Zhou, J. (eds.) WISTP 2011. LNCS, vol. 6633, pp. 224–233. Springer, Heidelberg (2011)
17. Zhao, V., Wang, T.: An Improved Differential Fault Attacks on Camellia. Cryptology ePrint Archive/585 (2009)
18. Zhao, X., Wang, T., Guo, S.: Further improved deep differential fault analysis on Camellia. In: The 2nd International Conference on Instrumentation, Measurement, Computer, Communication and Control - IMCCC, pp. 878–882, IEEE Computer Society (2012)

A Note on the Security of CHES 2014 Symmetric Infective Countermeasure

Alberto Battistello[1,2][✉] and Christophe Giraud[1]

[1] Cryptography and Security Group, Oberthur Technologies,
4, allée du Doyen Georges Brus, 33600 Pessac, France
{a.battistello,c.giraud}@oberthur.com
[2] Laboratoire de Mathématiques de Versailles, UVSQ, CNRS,
Université Paris-Saclay, 78035 Versailles, France

Abstract. Over the years, fault injection has become one of the most dangerous threats for embedded devices such as smartcards. It is thus mandatory for any embedded system to implement efficient protections against this hazard. Among the various countermeasures suggested so far, the idea of *infective computation* seems fascinating, probably due to its aggressive strategy. Originally conceived to protect asymmetric cryptosystems, infective computation has been recently adapted to symmetric systems. This paper investigates the security of a new symmetric infective countermeasure suggested at CHES 2014. By noticing that the number of executed rounds is not protected, we develop four different attacks that exploit the infection algorithm to disturb the round counter and related variables. Our attacks allow one to efficiently recover the secret key of the underlying cryptosystem by using any of the three most popular fault models used in literature.

Keywords: Fault attack · Infective countermeasure · AES

1 Introduction

Over the last 20 years, the security of embedded devices has been challenged by several specific attacks. In particular, Boneh *et al.* showed in 1996 that a simple disturbance during the execution of an embedded algorithm may totally break its security [5]. They illustrated this new method by explaining how to break an CRT-RSA implementation by inducing only one error during the algorithm execution. By using so-called *fault attacks*, many signature schemes and symmetric cryptosystems have been broken only a few months after the original Boneh *et al.* publication [1,4]. A whole new research field thus appeared aiming at discovering new fault-based attacks and providing efficient countermeasures [11,13,15]. While researchers improved and discovered new fault attacks on each and every cryptosystem, the countermeasures were difficult to find and costly to implement. Among the ideas that emerged, the two most popular methods are the signature verification for asymmetric systems and the duplication

© Springer International Publishing Switzerland 2016
F.-X. Standaert and E. Oswald (Eds.): COSADE 2016, LNCS 9689, pp. 144–159, 2016.
DOI: 10.1007/978-3-319-43283-0_9

method for symmetric ones. The first one simply consists in performing a signature verification on the result. If a fault occurred then the signature is not consistent and the verification fails. The second method requires to execute the algorithm twice and to compare both results. If an attacker disturbs one of the two executions then the comparison detects the attack and no output is returned. A third approach called *infective* was suggested in 2001 by Yen *et al.* [19]. Their method consists in modifying and amplifying the injected error in such a way that the attacker cannot retrieve any information from the corresponding faulty output. Firstly applied to asymmetric cryptosystems [19] this method is tricky to conceive and all infective countermeasures for asymmetric algorithms published so far have been broken, see [3,8,17,18] for instance. The infective method has been adapted only recently to the symmetric case. The first example of symmetric infection was proposed by Lomné *et al.* in 2012 to protect AES [12]. The authors suggest to execute the AES twice and to compute the infection by multiplicatively masking the differential of the two AES outputs. A second symmetric infection was suggested by Gierlichs *et al.* in [10] by using a random sequence of cipher and redundant rounds together with dummy rounds. If the outputs of the redundant and cipher rounds are different then the temporary result is infected. Unfortunately both methods have been broken by Battistello and Giraud in [2].

At CHES 2014, Tupsamudre *et al.* improved in [16] the attack of Battistello and Giraud and they also suggested an improved version of the infective countermeasure of Gierlichs *et al.* While this new proposal is secure against the attacks found in [2,16], one wonders if they are sufficient to make a symmetric implementation effectively secure, especially in the absence of a proof of security. Such a study has been done by Patranabis *et al.* in [14] where they provide an information theoretical analysis of the countermeasure suggested in [16]. They found weaknesses and proposed ways to reduce the efficiency of such threats.

In this paper, we extend the analysis of [14] by studying the security of the proposition of Tupsamudre *et al.* We firstly refine the attack presented in [14] and we analyze precisely its efficiency. We also suggest three other different attack paths that allow the attacker to modify the number of executed rounds by disturbing the infective algorithm variables. In order to mount our attacks we exploit three common fault models used in literature, from skip faults to random error faults. This paper not only shows that a straightforward implementation of the CHES 2014 infective countermeasure is insecure but also shows that implementers should pay particular attention to any aspect of a security countermeasure when implementing it.

The rest of the paper is organized as follows. In Sect. 2 we recall the countermeasure suggested in [16]. Section 3 presents four different attacks on this countermeasure. In particular, we show that it is possible to recover the secret key by using any of the three most popular fault models used in the literature. Section 4 finally concludes this paper.

2   Description of CHES 2014 Infective Countermeasure

The infective countermeasure suggested at CHES 2014 by Tupsamudre *et al.* [16] is based on the work presented at LatinCrypt 2012 by Gierlichs *et al.* [10]. For the sake of simplicity we recall in Algorithm 1 the CHES 2014 countermeasure suggested in [16] applied to AES-128. For more information about AES, the reader can refer to [9].

Algorithm 1. CHES 2014 COUNTERMEASURE APPLIED ON AES-128

Inputs : Plaintext P, round keys k^j for $j \in \{1, \ldots, 11\}$, pair (β, k^0), security
level $t \geq 22$

Output: Ciphertext $C = \text{AES-128}(P,K)$

1 State $R_0 \leftarrow P$; Redundant state $R_1 \leftarrow P$; Dummy state $R_2 \leftarrow \beta$
2 $i \leftarrow 1, q \leftarrow 1$
3 $rstr \leftarrow \{0,1\}^t$ // $\#1(rstr) = 22, \#0(rstr) = t - 22$
4 **while** $q \leq t$ **do**
5 $\quad \lambda \leftarrow rstr[q]$ // $\lambda = 0$ implies a dummy round
6 $\quad \kappa \leftarrow (i \wedge \lambda) \oplus 2(\neg\lambda)$
7 $\quad \zeta \leftarrow \lambda \cdot \lceil i/2 \rceil$ // ζ is actual round counter, $\qquad$ // $\zeta = 0$ is for
dummy round
8 $\quad R_\kappa \leftarrow \text{RoundFunction}(R_\kappa, k^\zeta)$
9 $\quad \gamma \leftarrow \lambda(\neg(i \wedge 1)) \cdot \text{BLFN}(R_0 \oplus R_1)$
10 $\quad \delta \leftarrow (\neg\lambda) \cdot \text{BLFN}(R_2 \oplus \beta)$
11 $\quad R_0 \leftarrow (\neg(\gamma \vee \delta) \cdot R_0) \oplus ((\gamma \vee \delta) \cdot R_2)$
12 $\quad i \leftarrow i + \lambda$
13 $\quad q \leftarrow q + 1$
14 **end**
15 **return** (R_0)

Algorithm 1 uses three states denoted R_0, R_1 and R_2 for the cipher, the redundant and the dummy rounds respectively. The execution order of these rounds is given by a random bit string $rstr$ generated at the beginning of the algorithm. Each "0" on the string encodes a dummy round, while a "1" encodes a redundant or cipher round. Each time a "1" occurs, an index i is incremented and a redundant round (resp. a cipher round) is executed if i is odd (resp. even). Algorithm 1 thus executes a loop over the $rstr$ string bits and executes a cipher, redundant or dummy round accordingly. One may note that Algorithm 1 computes the redundant round before the cipher round all along the algorithm and dummy rounds can happen randomly at any time.

Dummy rounds are executed over a dummy state R_2 which is initialized to a random 128-bit value β and by using the round key k^0 which is computed such that:

$$\text{RoundFunction}(\beta, k^0) = \beta. \tag{1}$$

The number of dummy rounds is parameterized by a security level t chosen by the developer. More precisely, this parameter represents the whole number of cipher, redundant and dummy rounds performed during Algorithm 1 execution. For instance in the case of AES-128, $t - 22$ dummy rounds will be performed.

Regarding the infective part, a first infection is activated after each cipher round if its state R_0 is different from the redundant state R_1 (Steps 9 and 11). Moreover another infection occurs if $R_2 \neq \beta$ after the execution of a dummy round (Steps 10 and 11). These infections consist in replacing the cipher state R_0 with the random value R_2, leaving no chance to the attacker to obtain information on the secret key once the infection is applied. To do so, a Boolean function BLFN is used which maps non-zero 128-bit values to 1 and outputs 0 for a null input.

Compared to the original LatinCrypt 2012 proposal, the CHES 2014 infective countermeasure differs by the way of dealing with the sequence of cipher, redundant and dummy rounds which is now done by using a random string $rstr$ and by the way the infection is performed which is now fully random.

Despite the security analysis of Algorithm 1 presented in [16], we show in the next section that it may be insecure if implemented as such. Furthermore we show that an attacker can recover the full secret key for each of the three most popular fault models used in literature.

3   Attacks

In this section we firstly present the principle of our attacks which are based on the fact that the variables dealing with the number of rounds to perform are not protected. We then exploit this remark to suggest four different attacks that use different fault models such as the instruction skip, the stuck-at and the random error fault model.

3.1   Principle of Our Attacks

Due to the improvements of Algorithm 1 compared with the original LatinCrypt 2012 countermeasure, it is impossible for the attacker to obtain any information on the secret key once the infection has occurred. In order to thwart this countermeasure, we thus investigate the possibility to disturb the number of executed rounds since the corresponding variable is not protected in integrity. Indeed, if the attacker succeeds in disturbing the number of rounds she may be able to retrieve the secret key from the corresponding faulty ciphertext [6,7]. In the remainder of this section, we show how such an attack works if the last round of an AES-128 has been skipped.

If the attacker knows a correct and a faulty ciphertext obtained by skipping the last AES round then it is equivalent to know the input S^{10} and the output S^{11} of the last round. Due to the lack of *MixColumns* transformation during the last AES round, the last round key k^{11} can be recovered byte per byte by XORing the corresponding bytes of S^{10} and S^{11}:

$$k_i^{11} = S_i^{11} \oplus \text{SBox}(S_{\text{SR}^{-1}(i)}^{10}), \quad \forall i \in \{1, \dots, 16\}, \tag{2}$$

where SR corresponds to the byte index permutation induced by the transformation *ShiftRows*. In such a case, the attacker can recover the full AES-128 key from only one pair of correct and faulty ciphertexts.

One may note that this attack works similarly if the attacker knows the input and the output of the first round. For further details on the first round attack the reader can refer to [6].

In the following, we describe different ways of disturbing Algorithm 1 by using several fault models such that it does not perform the AES with the correct number of rounds whereas no infection is performed. In our description we make use of AES-128 as a reference, however our attacks can apply straightforwardly to other key sizes.

3.2   Attack 1 by Using Instruction Skip Fault Model

The first attack that we present is an extension of the one presented in [14] which exploits the instruction skip fault model. The attack essentially works because whenever the variable i is odd and $\lambda = 1$ then a redundant round is executed and this kind of round does not involve any infection. In the following we assume that the attacker can skip an instruction of its choice by means of a fault injection.

Description. If the attacker skips Step 12 of Algorithm 1 after the last redundant round then the increment of i is not performed. Therefore i stays odd so the last cipher round is replaced by another redundant round. As no infection is involved for redundant rounds, the algorithm returns the output of the penultimate round. The attacker can thus take advantage of such an output to recover the secret key as explained in Sect. 3.1.

Efficiency. As explained in Appendix A, the probability of skipping the last cipher round and thus to recover the AES key after disturbing r different AES executions by skipping Step 12 during the q-th loop is given by:

$$\mathcal{P}_r = 1 - \left(1 - \frac{\binom{q-1}{20}\binom{t-q}{1}}{\binom{t}{22}} \right)^r. \tag{3}$$

where t is the total number of rounds performed during Algorithm 1, i.e. the number of while loops.

Some numerical values of (3) are given in Table 1 for t equal to 30, 40 and 50, $q = t - 3, \cdots, t - 1$ and $r = 1, \cdots, 4$. One can notice that if the fault is injected when q equals t then the attack does not work because all the rounds have already been executed.

By analyzing Table 1, one can deduce the best strategy for the attacker. For example if $t = 30$ then the attacker should target the 29-th loop in order to obtain the best chances of retrieving the key with the minimal number of fault injections.

Experiments. The attack described in this section has been simulated for $t = 30$ and for each q between 25 and 29. The experiment has been repeated 3 000 times for each configuration. The results of our tests are depicted in Fig. 1.

Table 1. Probability of obtaining at least one useful faulty ciphertext by skipping Step 12 during the q-th loop of Algorithm 1.

t	q	Number r of faults			
		1	2	3	4
30	27	11.80 %	22.21 %	31.39 %	39.49 %
	28	30.34 %	51.48 %	66.20 %	76.46 %
	29	53.10 %	78.01 %	89.69 %	95.16 %
40	37	19.34 %	34.93 %	47.52 %	57.66 %
	38	28.06 %	48.24 %	62.76 %	73.21 %
	39	29.62 %	50.46 %	65.13 %	75.46 %
50	47	18.96 %	34.32 %	46.77 %	56.86 %
	48	22.00 %	39.16 %	52.54 %	62.98 %
	49	18.86 %	34.16 %	46.57 %	56.65 %

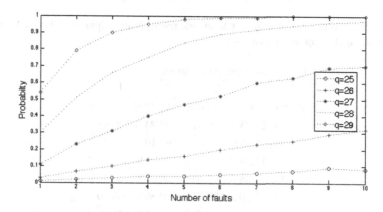

Fig. 1. Experimental probability of obtaining a useful faulty ciphertext by skipping Step 12 during the q-th loop of Algorithm 1 for $t = 30$.

By comparing Fig. 1 and the row $t = 30$ of Table 1, one can notice that the experiments perfectly match with the theoretical results.

3.3 Attack 2 by Using Stuck-At 0 Fault Model

In this section we use the *stuck-at 0* fault model where we assume that the attacker can set to zero a variable of her choice. As for the attack presented in Sect. 3.2, the goal of the attacker is to skip the execution of the last cipher round.

Description. To avoid the execution of the last cipher round by using a stuck-at 0 fault model without activating an infection, the attacker can set to zero the variable λ right after Step 5 during the loop involving the last "1" of *rstr*, i.e. during the loop dealing with the last cipher round. The computation of the last cipher round is thus skipped since $\lambda = 0$ implies a dummy round. The attacker thus retrieves an exploitable faulty ciphertext that can be used to retrieve the secret key as described in Sect. 3.1. As no consistency check is performed on λ, *rstr* nor on the number of cipher rounds executed, Algorithm 1 does not detect the fault.

Efficiency. We detail in Appendix B the reasoning to compute the probability of obtaining at least one useful faulty ciphertext after disturbing r different AES executions by setting λ to 0 after Step 5 of the q-th loop. Such a probability is given by:

$$\mathcal{P}_r = 1 - \left(1 - \frac{\binom{q-1}{21}}{\binom{t}{22}}\right)^r. \tag{4}$$

Some numerical values of (4) are given in Table 2 for t equal to 30, 40 and 50, $q = t - 2, \cdots, t$ and $r = 1, \cdots, 4$.

Table 2. Probability of obtaining at least one useful faulty ciphertext by sticking λ at 0 during the q-th loop of Algorithm 1

t	q	Number r of faults			
		1	2	3	4
30	28	5.06 %	9.86 %	14.42 %	18.75 %
	29	20.23 %	36.37 %	49.24 %	59.51 %
	30	73.33 %	92.89 %	98.10 %	99.49 %
40	38	11.36 %	21.42 %	30.35 %	38.26 %
	39	25.38 %	44.33 %	58.46 %	69.00 %
	40	55.00 %	79.75 %	90.89 %	95.90 %
50	48	14.14 %	26.29 %	36.71 %	45.66 %
	49	25.14 %	43.96 %	58.05 %	68.60 %
	50	44.00 %	68.64 %	82.44 %	90.17 %

By comparing Table 2 with Table 1, one may note that the attack presented in this section is more efficient than the one presented in Sect. 3.2, especially when the attacker targets the last loop execution.

Experiments. We simulated the attack for $t = 30$ and for q from 27 to 30. For each value of q we performed 3 000 tests with random *rstr*. The results of such experiments are depicted in Fig. 2.

3.4   Attack 3 by Using Random Error Fault Model

We show in this section how the attacker can use the random error fault model to obtain a useful faulty ciphertext. In this fault model, we assume that the attacker can change the value of a chosen internal variable into a random value.

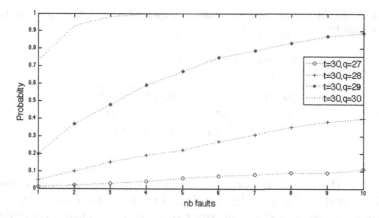

Fig. 2. Experimental probability of obtaining a useful faulty ciphertext by sticking λ at 0 during the q-th loop of Algorithm 1 for $t = 30$.

Description. Due to its central role in the infection and scheduling, string $rstr$ is very sensitive. However, the authors of [16] do not suggest any mean of ensuring its integrity. We thus investigated this path and we noticed that an attacker can disturb the generation of $rstr$ at Step 3 of Algorithm 1 such that it does not contain 22 "1" anymore. If the fault disturbs the string $rstr$ such that it contains only 21 (resp. 20) "1" then Algorithm 1 does not execute the last cipher round (resp. the last redundant and cipher rounds). In both cases no infection is performed allowing the attacker to exploit the corresponding faulty ciphertext to recover the secret key as explained in Sect. 3.1.

Efficiency. The probability to obtain at least one useful faulty ciphertext after disturbing r different AES executions by randomly modifying the least significant byte of $rstr$ during Step 3 is given by:

$$\mathcal{P}_r = 1 - \left(1 - \left(\sum_{i=1}^{8} \frac{\binom{t-8}{22-i}\binom{8}{i}}{\binom{t}{22}} \sum_{j=1}^{i} \frac{\binom{i}{j}\binom{8-i}{j-1}}{255} + \sum_{i=2}^{8} \frac{\binom{t-8}{22-i}\binom{8}{i}}{\binom{t}{22}} \sum_{j=2}^{i} \frac{\binom{i}{j}\binom{8-i}{j-2}}{255}\right)\right)^r. \tag{5}$$

For more details about the computation of this probability, the reader can refer to Appendix C.

Table 3 gives the probability to obtain a useful faulty ciphertext for t equal to 30, 40 and 50.

Experiments. Figure 3 shows the results obtained by simulating the attack described above. The simulations have been performed by generating a random string $rstr$ and disturbing it with an 8-bit random error. The test has been performed 3 000 times for each t equal to 30, 40 and 50.

3.5 Attack 4 by Using Random Error Fault Model

This section describes a second attack that can be mounted by using the random error fault model.

Table 3. Probability of obtaining at least one useful faulty ciphertext by disturbing Step 3 of Algorithm 1.

t	Number r of faults			
	1	2	3	4
30	41.63 %	65.93 %	80.11 %	88.39 %
40	34.72 %	57.39 %	72.18 %	81.84 %
50	24.60 %	43.15 %	57.13 %	67.67 %

Description. The idea of the attack is to disturb the increment of index q at Step 13 of Algorithm 1 during the execution of the first cipher round. We noticed that if the disturbance produces an error e such that $q \oplus e > t$ then the evaluation at Step 4 is false and the algorithm returns. If the algorithm computes only one cipher round then the attacker can use such an output to retrieve the first round key, cf. [6]. It is important to notice that in order to retrieve a useful output, the attacker needs to disturb the execution during the first cipher round and not after a redundant or dummy round.

Efficiency. As detailed in Appendix D, the probability to obtain at least one useful faulty ciphertext after disturbing r different AES executions by injecting a random error during Step 13 of the q-th loop is given by:

$$\mathcal{P}_r = 1 - \left(1 - \frac{2^8 - t}{2^8} \sum_{i=2}^{3} \frac{\binom{q}{i}\binom{t-q}{22-i}}{\binom{t}{22}}\right)^r. \tag{6}$$

We give in Table 4 the probability that the attacker retrieves a useful faulty ciphertext for t equal to 30, 40 and 50 and for q from 2 to 4.

The attacker can use Table 4 to choose the best strategy for her attack. For example for $t = 30$, one obtains the best chances to retrieve a useful faulty cipher-

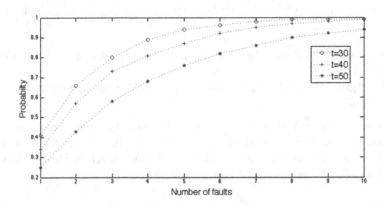

Fig. 3. Experimental probability of obtaining a useful faulty ciphertext by disturbing Step 3 of Algorithm 1.

Table 4. Probability of obtaining a useful faulty ciphertext by injecting a random error fault on Step 13 of Algorithm 1.

t	q	Number of faults			
		1	2	3	4
30	2	46.88 %	71.78 %	85.01 %	92.04 %
	3	73.67 %	93.07 %	98.17 %	99.52 %
	4	60.52 %	84.42 %	93.85 %	97.57 %
40	2	24.99 %	43.73 %	57.79 %	68.34 %
	3	48.66 %	73.64 %	86.47 %	93.05 %
	4	58.22 %	82.55 %	92.71 %	96.95 %
50	2	15.17 %	28.05 %	38.96 %	48.23 %
	3	32.88 %	54.95 %	69.76 %	79.70 %
	4	45.58 %	70.38 %	83.88 %	91.23 %

text by attacking the third loop. Furthermore when comparing the efficiency of our four attacks, the attack presented in this section is the most efficient one.

Experiments. We mounted several simulations where we disturbed the Step 13 of the q-th loop with a random byte error e. We mounted the experiments for $t = 30$ and for q from 2 to 6. For each different q we repeated the experiment 3 000 times. The results of such experiments are shown in Fig. 4.

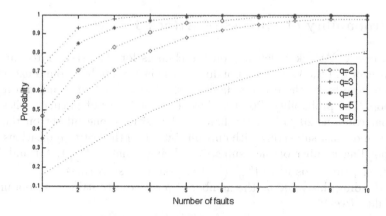

Fig. 4. Experimental probability of obtaining a useful faulty ciphertext by a injecting random error fault on Step 13 of Algorithm 1 for $t = 30$.

The simulations shows that this attack has a remarkable success rate. For example for $t = 30$, an attacker that reiterates the fault injection only twice during the third loop has a probability of retrieving a useful faulty ciphertext greater than 90 %.

4   Conclusion

In this article we showed that the infective countermeasure of CHES 2014 is not as secure as expected. While the countermeasure gives no information to the attacker once the infection is applied, we discovered that it does not protect the number of cipher rounds effectively executed. Despite the fact that attacks on the round counter are well known, our work describes attack paths that are difficult to spot and involve disturbances on the infective variables intentionally added to thwart fault attacks. The aim of this paper is thus to warn the reader of possible security weaknesses that may reside in straightforward implementations of the countermeasure.

We applied the three most popular fault models and found four different attack paths that allow an attacker to recover the secret key of the underlying cryptosystem. For each attack we studied the success probability and performed simulations that validated our theoretical results.

An obvious countermeasure consists in ensuring the integrity of i, q, λ and $rstr$ for instance. In their work Patranabis et $al.$ [14] suggest a possible countermeasure based on this remark to thwart the instruction skip fault model. However, their analysis does not take into account other fault models that are exploited in this work. We thus suggest that a possible idea for future improvements may be to fill this gap.

With this work we also remark that the lack of formal security proofs in this field is clearly an issue. We hope that new ideas may pave the way to formally prove the security of cryptosystems against fault-based cryptanalysis.

A   Probability of Success of Attack 1

The success of Attack 1 depends on the chances for the attacker to fault the increment of i in the loop corresponding to the last redundant round execution. Let us denote by e_1 the event of faulting the last redundant round during the q-th loop. The probability $\mathcal{P}(e_1)$ is thus the probability of having a bit-string $rstr$ that contains 20 "1" on the first $q-1$ positions, one bit set on the q-th position and a last sub-string with only one bit set on the last $t-q$ positions. The corresponding number of such sub-strings being equal to $\binom{q-1}{20}$, $\binom{1}{1}$ and $\binom{t-q}{1}$ respectively, this leads us to $\binom{q-1}{20}\binom{t-q}{1}$ exploitable $rstr$ strings.

By dividing this value by the number of possible $rstr$ strings, we obtain the probability $\mathcal{P}(e_1)$:

$$\mathcal{P}(e_1) = \frac{\binom{q-1}{20}\binom{t-q}{1}}{\binom{t}{22}}. \tag{7}$$

As described in Appendix E, we then compute by using Eq. (20) the probability to obtain at least one useful faulty ciphertext by repeating the fault injection r times.

B   Probability of Success of Attack 2

Let us evaluate the probability that the event e_2 of obtaining a useful faulty ciphertext by setting to zero the variable λ at Step 5 of Algorithm 1 happens. The probability $\mathcal{P}(e_2)$ corresponds to the probability of obtaining a string $rstr$ that has 21 bits set on the first $q-1$ positions, a "1" on the q-th position and only "0"'s on the last $t - q$ positions. As we have done in Appendix A, we compute this probability as the number of such strings divided by the total number of possible $rstr$ strings. As there is only one possibility that the last $t - (q - 1)$ bits of $rstr$ are exactly "$10\cdots0$", we thus obtain:

$$\mathcal{P}(e_2) = \frac{\binom{q-1}{21}}{\binom{t}{22}}, \tag{8}$$

As described in Appendix E, we then compute by using Eq. (20) the probability to obtain at least one useful faulty ciphertext by repeating the fault injection r times.

C   Probability of Success of Attack 3

Let us denote by e_3 the event that a random byte error disturbs the string $rstr$ such that it contains only 21 or 20 "1". To evaluate the probability $\mathcal{P}(e_3)$ that the event e_3 occurs, let us assume for the sake of simplicity that the attacker disturbs the least significant byte B of $rstr$ which corresponds to a random byte fault model. By firstly evaluating the case 21, we observe that the probability that a bit-string has exactly 21 bits set on the first $t - 8$ positions and the remaining "1" in one of the last 8 positions is:

$$\mathcal{P}(HW(B) = 1) = \frac{\binom{t-8}{21}\binom{8}{1}}{\binom{t}{22}}, \tag{9}$$

where we denote by $HW(B)$ the Hamming weight of the byte B. Equation (9) corresponds to the probability that the last byte of $rstr$ has an Hamming weight equal to 1. By summing the corresponding probabilities for all the Hamming weights between 1 and 8 we obtain the probability that the last byte of $rstr$ has an Hamming weight greater than zero:

$$\mathcal{P}(HW(B) > 0) = \sum_{i=1}^{8} \frac{\binom{t-8}{22-i}\binom{8}{i}}{\binom{t}{22}}. \tag{10}$$

Now, let us compute the probability of injecting a random error on a byte of Hamming weight i such that the byte contains only $i-1$ "1" after the disturbance. We thus count for each possible value of B how many 8-bit values e exist such that $HW(B \oplus e) = HW(B) - 1$. This corresponds to the number of possible errors setting to "0" j bits "1" while setting to "1" $j-1$ bits "0". Afterwards we divide the result by the number of possible values for the error e:

$$P(HW(B \oplus e) = HW(B) - 1|B)$$
$$= \frac{\sum_{j=1}^{HW(B)} \binom{HW(B)}{j} \binom{8-HW(B)}{j-1}}{255}. \qquad (11)$$

This corresponds to the probability that $HW(B \oplus e) = HW(B) - 1$ by injecting a random error e on a random 8-bit value B.

By combining the two probabilities above, we obtain the probability that $rstr$ contains 21 "1" after a random error injection on the last byte of $rstr$:

$$P(HW(B \oplus e) = 21) = \sum_{i=1}^{8} \frac{\binom{t-8}{22-i}\binom{8}{i}}{\binom{t}{22}} \sum_{j=1}^{i} \frac{\binom{i}{j}\binom{8-i}{j-1}}{255}. \qquad (12)$$

For the case where $rstr$ contains only 20 "1", we use the same reasoning and we obtain:

$$P(HW(B \oplus e) = 20) = \sum_{i=2}^{8} \frac{\binom{t-8}{22-i}\binom{8}{i}}{\binom{t}{22}} \sum_{j=2}^{i} \frac{\binom{i}{j}\binom{8-i}{j-2}}{255}. \qquad (13)$$

Thus the total probability of disturbing the generation of one byte of $rstr$ such that it contains a total of 21 or 20 "1" is:

$$P(e_3) = \sum_{i=1}^{8} \frac{\binom{t-8}{22-i}\binom{8}{i}}{\binom{t}{22}} \sum_{j=1}^{i} \frac{\binom{i}{j}\binom{8-i}{j-1}}{255} + \sum_{i=2}^{8} \frac{\binom{t-8}{22-i}\binom{8}{i}}{\binom{t}{22}} \sum_{j=2}^{i} \frac{\binom{i}{j}\binom{8-i}{j-2}}{255}. \qquad (14)$$

As described in Appendix E, we then compute by using Eq. (20) the probability to obtain at least one useful faulty ciphertext by repeating the fault injection r times.

D   Probability of Success of Attack 4

In the following we denote by e_4 the event that the error e is injected after a cipher round and is such that $q \oplus e > t$. In order to evaluate the probability $P(e_4)$ we need to compute:

- the probability that the error e leads to $q \oplus e > t$,
- the probability that the attacker disturbs the algorithm after a cipher round and not after a redundant or dummy round.

For the first probability, without loss of generality, we assume that q is coded over one byte which should be the case in practice. We thus obtain that the probability of injecting an 8-bit error e such that $q \oplus e > t$ depends only on t and is given by:

$$P(q \oplus e > t) = \frac{2^8 - t}{2^8}. \qquad (15)$$

In order to evaluate the second probability we remark that it is equivalent to the probability that the string $rstr$ contains two or three "1" in the first q positions. We recall that $rstr$ is a string with 22 "1" at most. Thus the number of possible strings $rstr$ with only two "1" in the first q positions is:

$$\binom{q}{2}\binom{t-q}{20}. \tag{16}$$

Summing Eq. (16) to the number of possible strings $rstr$ with only three "1" in the first q positions we obtain the number of favorable cases for the attacker:

$$\binom{q}{2}\binom{t-q}{20} + \binom{q}{3}\binom{t-q}{22-3}. \tag{17}$$

By dividing by the total number of possible $rstr$ strings we thus obtain the probability that the algorithm has executed only one cipher round after q rounds:

$$P(HW(rstr[1,\ldots,q]) \in [2,3]) = \frac{\binom{q}{2}\binom{t-q}{20} + \binom{q}{3}\binom{t-q}{19}}{\binom{t}{22}}, \tag{18}$$

where $rstr[1,\ldots,q]$ denotes the sub-string of $rstr$ between the first and the q-th position. By combining the two probabilities we obtain:

$$P(e_4) = \frac{2^8 - t}{2^8} \sum_{i=2}^{3} \frac{\binom{q}{i}\binom{t-q}{22-i}}{\binom{t}{22}}, \tag{19}$$

which corresponds to the probability that the algorithm returns an exploitable faulty ciphertext by injecting a random error after q rounds.

As described in Appendix E, we then compute by using Eq. 20 the probability to obtain at least one useful faulty ciphertext by repeating the fault injection r times.

E   Attack Repetition Probability

For each attack, we denote by $P(e_i)$ the probability that event e_i occurs. By assuming that $P(e_i)$ is independent for each execution we can compute the probability of getting at least one useful faulty ciphertext by repeating the fault injection r times as:

$$P_r = 1 - (1 - P(e_i))^r. \tag{20}$$

References

1. Bao, F., Deng, R., Han, Y., Jeng, A., Narasimhalu, A.D., Ngair, T.-H.: Breaking public key cryptosystems and tamper resistance devices in the presence of transient fault. In: Christianson, B., Crispo, B., Lomas, M., Roe, M. (eds.) Security Protocols 1997. LNCS, vol. 1361, pp. 115–124. Springer, Heidelberg (1998)

2. Battistello, A., Giraud, C.: Fault analysis of infective AES computations. In: Fischer, W., Schmidt, J.-M. (eds.) FDTC, pp. 101–107. IEEE (2013)
3. Berzati, A., Canovas, C., Goubin, L.: (In)security against fault injection attacks for CRT-RSA implementations. In: Breveglieri, L., Gueron, S., Koren, I., Naccache, D., Seifert, J.-P. (eds.) Fault Diagnosis and Tolerance in Cryptography - FDTC, pp. 101–107. IEEE Computer Society (2008)
4. Biham, E., Shamir, A.: Differential fault analysis of secret key cryptosystems. In: Kaliski Jr., B.S. (ed.) CRYPTO 1997. LNCS, vol. 1294, pp. 513–525. Springer, Heidelberg (1997)
5. Boneh, D., DeMillo, R.A., Lipton, R.J.: On the importance of checking cryptographic protocols for faults. In: Fumy, W. (ed.) EUROCRYPT 1997. LNCS, vol. 1233, pp. 37–51. Springer, Heidelberg (1997)
6. Choukri, H., Tunstall, M.: Round reduction using faults. In: Breveglieri, L., Koren, I. (eds.) Workshop on Fault Diagnosis and Tolerance in Cryptography - FDTC (2005)
7. Dutertre, J.-M., Mirbaha, A.-P., Naccache, D., Ribotta, A.-L., Tria, A., Vaschalde, T.: Fault round modification analysis of the advanced encryption standard. In: IEEE International Symposium on Hardware-Oriented Security and Trust - HOST, pp. 28–39. IEEE (2012)
8. Feix, B., Venelli, A.: Defeating with fault injection a combined attack resistant exponentiation. In: Prouff, E. (ed.) COSADE 2013. LNCS, vol. 7864, pp. 32–45. Springer, Heidelberg (2013)
9. FIPS PUB 197. Advanced Encryption Standard. National Institute of Standards and Technology, November 2001
10. Gierlichs, B., Schmidt, J.M., Tunstall, M.: Infective computation and dummy rounds: fault protection for block ciphers without check-before-output. In: Hevia, A., Neven, G. (eds.) LatinCrypt 2012. LNCS, vol. 7533, pp. 305–321. Springer, Heidelberg (2012)
11. Giraud, C.: DFA on AES. In: Dobbertin, H., Rijmen, V., Sowa, A. (eds.) AES 2005. LNCS, vol. 3373, pp. 27–41. Springer, Heidelberg (2005)
12. Lomné, V., Roche, T., Thillard, A.: On the need of randomness in fault attack countermeasures - application to AES. In: Bertoni, G., Gierlichs, B. (eds.) Fault Diagnosis and Tolerance in Cryptography - FDTC, pp. 85–94. IEEE Computer Society (2012)
13. Mukhopadhyay, D.: An improved fault based attack of the advanced encryption standard. In: Preneel, B. (ed.) AFRICACRYPT 2009. LNCS, vol. 5580, pp. 421–434. Springer, Heidelberg (2009)
14. Patranabis, S., Chakraborty, A., Mukhopadhyay, D.: Fault Tolerant Infective Countermeasure for AES. Cryptology ePrint Archive, Report 2015/493 (2015). http://eprint.iacr.org/
15. Piret, G., Quisquater, J.J.: A differential fault attack technique against SPN structures, with application to the AES and KHAZAD. In: Walter, C.D., Koç, Ç.K., Paar, C. (eds.) CHES 2003. LNCS, vol. 2779, pp. 77–88. Springer, Heidelberg (2003)
16. Tupsamudre, H., Bisht, S., Mukhopadhyay, D.: Destroying fault invariant with randomization. In: Batina, L., Robshaw, M. (eds.) CHES 2014. LNCS, vol. 8731, pp. 93–111. Springer, Heidelberg (2014)
17. Wagner, D.: Cryptanalysis of a provable secure CRT-RSA algorithm. In: Pfitzmann, B., Liu, P. (eds.) ACM Conference on Computer and Communications Security - CCS 2004, pp. 82–91. ACM Press (2004)

18. Yen, S.M., Kim, D., Moon, S.J.: Cryptanalysis of two protocols for RSA with CRT based on fault infection. In: Breveglieri, L., Koren, I., Naccache, D., Seifert, J.P. (eds.) FDTC 2006. LNCS, vol. 4236, pp. 53–61. Springer, Heidelberg (2006)
19. Yen, S.-M., Kim, S., Lim, S., Moon, S.-J.: RSA speedup with residue number system immune against hardware fault cryptanalysis. In: Kim, K. (ed.) ICISC 2001. LNCS, vol. 2288, pp. 397–413. Springer, Heidelberg (2002)

Side-Channel Analysis (Tools)

Simpler, Faster, and More Robust T-Test Based Leakage Detection

A. Adam Ding[1], Cong Chen[2(✉)], and Thomas Eisenbarth[2]

[1] Northeastern University, Boston, MA, USA
a.ding@neu.edu
[2] Worcester Polytechnic Institute, Worcester, MA, USA
{cchen3,teisenbarth}@wpi.edu

Abstract. The TVLA procedure using the t-test has become a popular leakage detection method. To protect against environmental fluctuation in laboratory measurements, we propose a paired t-test to improve the standard procedure. We take advantage of statistical matched-pairs design to remove the environmental noise effect in leakage detection. Higher order leakage detection is further improved with a moving average method. We compare the proposed test with standard t-test on synthetic data and physical measurements. Our results show that the proposed tests are robust to environmental noise.

1 Motivation

More than 15 years after the proposal of DPA, standardized side channel leakage detection is still a topic of controversial discussion. While Common Criteria (CC) testing is an established process for highly security critical applications such as banking smart cards and passport ICs, the process is slow and costly. While appropriate for high-security applications, CC is too expensive and too slow to keep up with the innovation cycle of a myriad of new networked embedded products that are currently being deployed as the Internet of Things. As a result, an increasing part of the world we live in will be monitored and controlled by embedded computing platforms that, without the right requirements in place, will be vulnerable to even the most basic physical attacks such as straightforward DPA.

A one-size-fits-most leakage detection test that is usable by non-experts and can reliably distinguish reasonably-well protected cryptographic implementations from insecure ones could remedy this problem. Such a test would allow industry to self-test their solutions and hopefully result in a much broader deployment of appropriately protected embedded consumer devices. The TVLA test was proposed as such a leakage detection test in [6,10]. The TVLA test checks if an application behaves differently under two differing inputs, e.g. one fixed input vs. one random input. As the original DPA, it uses averaging over a large set of observations to detect even most nimble differences in behavior, which can potentially be exploited by an attacker.

© Springer International Publishing Switzerland 2016
F.-X. Standaert and E. Oswald (Eds.): COSADE 2016, LNCS 9689, pp. 163–183, 2016.
DOI: 10.1007/978-3-319-43283-0_10

Due to its simplicity, it is applicable to a fairly wide range of cryptographic implementations. In fact, academics have started to adopt this test to provide evidence of existing leakages or their absence [1,3–5,13,15,16,20]. With increased popularity, scrutiny of the TVLA test has also increased. Mather et al. [14] studied the statistical power and computation complexity of the t-test versus mutual information (MI) test, and found that t-test does better in the majority of cases. Schneider and Moradi [19] for example showed how the t-test higher order moments can be computed in a single pass. They also discussed the tests sensitivity to the measurement setup and proposed a randomized measurement order. Durveaux and Standaert [8] evaluate the convenience of the TVLA test for detecting relevant points in a leakage trace. They also uncover the implications of good and bad choices of the fixed case for the fixed-vs-random version of the TVLA test and discuss the potential of a fixed-vs-fixed scenario.

However, there are other issues besides the choice of the fixed input and the measurement setup that can negatively impact the outcome for the t-test based leakage detection. Environmental effects can influence the t-test in a negative way, i.e., will decrease its sensitivity. In the worst case, this means that a leaky device may pass the test only because the environmental noise was strong enough. This is a problem for the proposed objective of the TVLA test, i.e. self-certification by non-professionals who are not required to have a broad background in side channel analysis.

Our Contribution. In this work, we propose the adoption of the paired t-test for leakage detection, especially in cases where long measurement campaigns are performed to identify nimble leakages. We discuss several practical issues of the classic t-test used in leakage detection and show that many of them can be avoided when using the paired t-test. To reap the benefits of the locality of the individual differences of the paired t-test in the higher order case, we further propose to replace the centered moments with a local approximation. These approximated central moments are computed over a small and local moving window, making the entire process a single-pass analysis. In summary, we show that

- the paired t-test is more robust to environmental noise such as temperature changes and drifts often observed in longer measurement campaigns, resulting in a faster and more reliable leakage detection.
- using moving averages instead of a central average results in much better performance for higher order and multivariate leakage detection if common measurement noise between the two classes of traces is present, while introducing a vanishingly small inaccuracy if no such common noise appears. The improvement of the moving averages applies both to the paired and unpaired t-tests.

In summary, we advocate the adoption of the paired t-test based on moving averages as a replacement of Welch' t-test for detecting leakages, as results are at least on par with the prevailing methodology while showing much better results in the presence of a common noise source.

2   Background

In the framework of [10], the potential leakage for a device under test (DUT) can be detected by comparing two sets of measurements $\mathcal{L}_A$ and $\mathcal{L}_B$ on the DUT. A popular test for the comparison is Welch's t-test, which aims to detect the mean differences between the two sets of measurements. The null hypothesis is that the two samples come from the same population so that their population means μ_A and μ_B are the same. Let $\bar{L}_A$ and $\bar{L}_B$ denote their sample means, s_A^2 and s_B^2 denote their sample variance, n_A and n_B denote the number of measurements in each set. Then the t-test statistic and its degree of freedom are given by

$$t_u = \frac{\bar{L}_A - \bar{L}_B}{\sqrt{\frac{s_A^2}{n_A} + \frac{s_B^2}{n_B}}}, \qquad v = \frac{(\frac{s_A^2}{n_A} + \frac{s_B^2}{n_B})^2}{\frac{(\frac{s_A^2}{n_A})^2}{n_A - 1} + \frac{(\frac{s_B^2}{n_B})^2}{n_B - 1}}. \tag{1}$$

The p-value of the t-test is calculated as the probability, under a t-distribution with v degree of freedom, that the random variable exceeds observed statistic $|t_u|$. This is readily done in Matlab as $2 * (1 - tcdf(\cdot, v))$ and in R as $2 * (1 - qt(\cdot, df = v))$. The null hypothesis of no leakage is rejected when the p-value is smaller than a threshold, or equivalently when the t-test statistic $|t_u|$ exceeds a corresponding threshold. The rejection criterion of $|t_u| > 4.5$ is often used [10,19]. Since $Pr(|t_{df=v>1000}| > 4.5) < 0.00001$, this threshold leads to a confidence level > 0.99999.

For leakage detection, a *specific* t-test use two sets $\mathcal{L}_A$ and $\mathcal{L}_B$ corresponding to different values of an intermediate variable: $V = v_A$ and $V = v_B$. To avoid the dependence on the intermediate value and the power model, *non-specific* t-test often uses the *fixed versus random* setup. That is, the first set $\mathcal{L}_A$ is collected with a fixed plaintext x_A, while the second set $\mathcal{L}_B$ is collected with random plaintexts x_B drawn from the uniform distribution. Then if there is leakage through an (unspecified) intermediate variable V, then

$$L_A = V(k, x_A) + r_A \qquad L_B = V(k, x_B) + r_B, \tag{2}$$

where k is the secret key, r_A and r_B are random measurement noises with zero means and variance σ_A^2 and σ_B^2 respectively. The non-specific t-test can detect the leakage, with large numbers of measurements n_A and n_B, when the fixed intermediate state $V(k, x_A)$ differs from the expected value of the random intermediate state $E_{x_B}[V(k, x_B)]$ where the expectation is taken over the uniform random plaintexts x_B.

The power model is very general for t-test framework of [10]. The intermediate variable can be of various sizes, including one bit or one byte intermediate state. Particularly, the tester does not need to know the underlying power model for the unspecified t-test. The power model in most of the paper is kept abstract and general. The theory does not depend on any specific power model. We only specify the exact power model in simulation studies that generated the data.

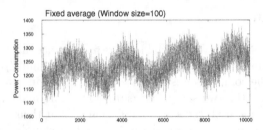

Fig. 1. Power consumption moving averages at a key-sensitive leakage point on the DPAv2 template traces

3 Methodology

This section introduces paired t-test and shows its superiority in a leakage model with environmental noise. The paired t-test retains its advantage of being a straightforward one-pass algorithm by making use of *moving* or local averages. By relying on the difference of matched pairs, the method is inherently numerically stable while retaining computational efficiency and parallelizability of the original t-test.

3.1 Paired T-Test

Welch's t-test works well when the measurement noises r_A and r_B are independent between the two sets of measurements. However, two sets of measurements can also share common variation sources during a measurement campaign. For example, power consumption and variance may change due to common environmental factors such as room temperature. While these environmental factors usually change slowly, such noise variation is more pronounced over a longer time period. With hard to detect leakages, often hundreds of thousands to millions of measurement traces are required for detection. These measurements usually take many hours and the environmental fluctuation is of concern in such situations. For example, for the DPA V2 contest, there are one million template traces collected over 3 days and 19 h, which show a clear temporal pattern [11]. Figure 1 (a subgraph of 2 in [11]) shows the average power consumption at 2373-th time point on the traces of DPAv2, using mean values over 100 non-overlapping subsequent traces.

Testing labs usually try to control the environmental factors to reduce such temporal variation. However, such effort can be expensive and there is no guarantee that all noise induced by environmental factors can be removed. Instead, we can deal with these environmental noise through statistical design. Particularly, we can adopt the *matched-pairs design* (Sect. 15.3 in [12]), where the measurements are taken in pairs with one each from the groups A and B. Then a *paired t-test* can be applied on such measurements, replacing the unpaired t-test (1). With n such pairs of measurements, we have n difference measurements $D = L_A - L_B$. The paired difference cancels the noise variation from the common source, making it easier to detect nonzero population difference. The null hypothesis of $\mu_A = \mu_B$ is equivalent to that the mean difference $\mu_D = 0$, which is tested by a paired t-test. Let $\bar{D}$ and s_D^2 denote the sample mean and sample variances of the paired differences $D_1, ..., D_n$. The paired t-test statistic is

$$t_p = \frac{\bar{D}}{\sqrt{\frac{s_D^2}{n}}}, \tag{3}$$

with the degree of freedom $n - 1$. The null hypothesis of non-leakage is rejected when $|t_p|$ exceeds the threshold of 4.5.

To quantify the difference between the two versions of t-test, we can compare the paired t-test (3) and the unpaired t-test (1) here with $n_A = n_B = n$.

First, without common variation sources under model (2), $Var(D) = Var(L_A) + Var(L_B) = \tilde{\sigma}_A^2 + \tilde{\sigma}_B^2$. Here $\tilde{\sigma}_A^2 = \sigma_A^2 + Var[V(k, x_A)]$ and $\tilde{\sigma}_B^2 = \sigma_B^2 + Var[V(k, x_B)]$. Notice that $\bar{D} = \bar{L}_A - \bar{L}_B$, so for large n, the paired t-test and unpaired t-test are equivalent with $t_u \approx t_p \approx (\bar{L}_A - \bar{L}_B)/\sqrt{(\tilde{\sigma}_A^2 + \tilde{\sigma}_B^2)/n}$. The paired t-test works even if the two group variances are unequal $\tilde{\sigma}_A^2 \neq \tilde{\sigma}_B^2$. The two versions of the t-test perform almost the same in this case.

However, the paired t-test detects leakage faster if there are common noise variation sources. To see this, we explicitly model the common environmental factor induced variation not covered by model (2).

$$L_A = V(k, x_A) + r_A + r_E \qquad L_B = V(k, x_B) + r_B + r_E, \tag{4}$$

where r_E is the noise caused by common environmental factors, with mean zero and variance σ_E. The r_A and r_B here denote the random measurement noises excluding common variations so that r_A and r_B are independent, with zero means and variance σ_A^2 and σ_B^2 respectively. Again we denote $\tilde{\sigma}_A^2 = \sigma_A^2 + Var[V(k, x_A)]$ and $\tilde{\sigma}_B^2 = \sigma_B^2 + Var[V(k, x_B)]$. Then $t_u \approx (\bar{L}_A - \bar{L}_B)/\sqrt{(\tilde{\sigma}_A^2 + \tilde{\sigma}_B^2 + 2\sigma_E^2)/n}$ while $t_p \approx (\bar{L}_A - \bar{L}_B)/\sqrt{(\tilde{\sigma}_A^2 + \tilde{\sigma}_B^2)/n}$. The paired t-test statistic $|t_p|$ has a bigger value than the unpaired t-test $|t_u|$, thus identifies the leakage more efficiently. The difference increases when the environmental noise σ_E increases. Hence, the paired t-test performs as well or better than the unpaired test. However, the matched-pairs design of the paired t-test cancels common noise found in both pairs, making the test more robust to suboptimal measurement setups and environmental noise.

3.2   Higher Order and Multivariate Leakage Detection

The t-test can also be applied to detect higher order leakage and multivariate leakage [10, 19]. For d-th order leakage at a single time point, the t-test compares sample means of $(L_A - \bar{L}_A)^d$ and $(L_B - \bar{L}_B)^d$. Under the matched-pairs design, the paired t-test would simply work on the difference

$$D = [(L_A - \bar{L}_A)^d - (L_B - \bar{L}_B)^d] \tag{5}$$

to yield the test statistic (3): $t_p = \bar{D}/\sqrt{s_D^2/n}$. Multivariate leakage combines leakage observation at multiple time points. A d-variate leakage combines leakage $L^{(1)}$, ..., $L^{(d)}$ at the d time points t_1, ..., t_d respectively. The combination is done through the centered product $CP(L^{(1)}, ..., L^{(d)}) = (L^{(1)} - \bar{L}^{(1)})(L^{(2)} - \bar{L}^{(2)})$ $\cdots (L^{(d)} - \bar{L}^{(d)})$. The standard d-variate leakage detection t-test compares the sample means of $CP(L_A^{(1)}, ..., L_A^{(d)})$ and $CP(L_B^{(1)}, ..., L_B^{(d)})$ with statistic (1). The paired t-test (3) uses the difference $D = [CP(L_A^{(1)}, ..., L_A^{(d)}) - CP(L_B^{(1)}, ..., L_B^{(d)})]$.

However, these tests (including the paired t-test) do not eliminate environmental noise effects on the higher order and multivariate leakage detection. The centering terms (the subtracted $\bar{L}$) in the combination function also need adjustment due to environmental noises, which are not random noise but follow some temporal patterns. To see this, we use the bivariate leakage model for first-order masked device as an example.

The leakage measurements at the two time points t_1 and t_2 leak two intermediate values $V^{(1)}(k, x, m)$ and $V^{(2)}(k, x, m)$ where k, x and m are the secret key, plaintext and mask respectively. For uniformly distributed m, $V^{(1)}(k, x, m)$ and $V^{(2)}(k, x, m)$ both follow a distribution not affected by k and x, therefore no first order leakage exits. Without loss of generality, we assume that $E_m[V^{(1)}(k, x, m)] = E_m[V^{(2)}(k, x, m)] = 0$, and the second order leakage comes from the product combination $V^{(1)}V^{(2)}$. [18] derived the strongest leakage combination function under a second order leakage model without the environmental noises:

$$L^{(1)} = c^{(1)} + V^{(1)}(k, x, m) + r^{(1)}, \qquad L^{(2)} = c^{(2)} + V^{(2)}(k, x, m) + r^{(2)}, \quad (6)$$

where $r^{(1)}$ and $r^{(2)}$ are zero-mean random pure measurement noises with variance σ_1^2 and σ_2^2 respectively. Under model (6), [18] showed that centered product leakage $(L^{(1)} - c^{(1)})(L^{(2)} - c^{(2)})$ is the strongest. Since c_1 and c_2 are unknown in practice, they are estimated by $\bar{L}^{(1)} = \bar{c}^{(1)} + \bar{V}^{(1)} + \bar{r}^{(1)}$ and $\bar{L}^{(2)} = \bar{c}^{(2)} + \bar{V}^{(2)} + \bar{r}^{(2)}$. With large number of traces, $\bar{L}^{(1)} \approx \bar{c}^{(1)}$ and $\bar{L}^{(2)} \approx \bar{c}^{(2)}$ by the law of large number. Hence $(L^{(1)} - \bar{L}^{(1)})(L^{(2)} - \bar{L}^{(2)})$ approximate the optimal leakage $(L^{(1)} - c^{(1)})(L^{(2)} - c^{(2)})$ well. However, considering environment induced noises, this is no longer the strongest leakage combination function. Let us assume that

$$L^{(1)} = c^{(1)} + V^{(1)}(k, x, m) + r^{(1)} + r_E^{(1)}, \; L^{(2)} = c^{(2)} + V^{(2)}(k, x, m) + r^{(2)} + r_E^{(2)}, \quad (7)$$

where $r_E^{(1)}$ and $r_E^{(2)}$ are environment induced noises which has mean zero but follow some temporal pattern rather than being random noise. The optimal leakage then becomes $(L^{(1)} - c^{(1)} - r_E^{(1)})(L^{(2)} - c^{(2)} - r_E^{(2)})$ instead. Therefore, we propose that the centering means $\bar{L}^{(1)}$ and $\bar{L}^{(2)}$ are calculated as moving averages from traces with a window of size n_w around the trace to be centered, rather than the average over all traces. The temporal patterns for $r_E^{(1)}$ and $r_E^{(2)}$, such as in Fig. 1, are usually slow changing. Hence, for a moderate window size, say $n_w = 100$, the moving averages $\bar{L}^{(1)} \approx c^{(1)} + r_E^{(1)}$ and $\bar{L}^{(2)} \approx c^{(2)} + r_E^{(2)}$.

When there are no environment induced noises $r_E^{(1)}$ and $r_E^{(2)}$, using bigger window size n_w can improve the precision. However, comparing to centering on averages of all traces, we can prove that centering the moving averages only loses $O(1/n_w)$ proportion of statistical efficiency under model (6). More precisely, denote the theoretical optimal leakage detection statistic as

$$\Delta = (L_A^{(1)} - c^{(1)})(L_A^{(2)} - c^{(2)}) - (L_B^{(1)} - c^{(1)})(L_B^{(2)} - c^{(2)}). \qquad (8)$$

And denote the leakage detection statistic using moving average of a window size n_w as

$$D = (L_A^{(1)} - \bar{L}_A^{(1)})(L_A^{(2)} - \bar{L}_A^{(2)}) - (L_B^{(1)} - \bar{L}_B^{(1)})(L_B^{(2)} - \bar{L}_B^{(2)}). \qquad (9)$$

Then for large sample size n, the t-test statistic (3) is approximately $t_p(D) \approx E(D)/\sqrt{Var(D)/n}$, and the optimal leakage detection t-test statistic is approximately $t_p(\Delta) \approx E(\Delta)/\sqrt{Var(\Delta)/n}$. A quantitative comparison of these two statistic is given in the next Theorem.

Theorem 1. *Under the second-order leakage model (6),*

$$\frac{E(D)}{\sqrt{Var(D)/n}} \frac{\sqrt{Var(\Delta)/n}}{E(\Delta)} = 1 - \frac{\eta}{n_w} + O(\frac{1}{n_w^2}), \qquad (10)$$

where the factor η is given by

$$\eta = \frac{1}{Var(\Delta)}[Var(V_A^{(1)})Var(V_A^{(2)}) + Var(V_B^{(1)})Var(V_B^{(2)}) + E^2(V_A^{(1)}V_A^{(2)})$$
$$+ E^2(V_B^{(1)}V_B^{(2)}) - Var(V_A^{(1)}V_A^{(2)}) - Var(V_B^{(1)}V_B^{(2)})].$$

The proof of Theorem 1 is provided in Appendix A.

The factor η is usually small. When the noise variances σ_1^2 and σ_2^2 are big (so that the leakage is hard to detect), this factor $\eta = O[1/(\sigma_1^2\sigma_2^2)] \approx 0$. For practical situations, often $\eta < 1$. Hence using, say, $n_w = 100$ make the leakage detection statistic robust to environmental noises $r_E^{(1)}$ and $r_E^{(2)}$, at the price of a very small statistical efficiency loss when no environmental noises exist. Therefore, we recommend this paired moving-average based t-test (MA-t-test) over the existing tests.

We can also estimate the optimal window size n_w with some rough ideas of environmental noise fluctuation. The potential harm in using too wide a window is to introduce bias in the estimated centering quantities. Let the environmental noise be described as $r_E(t)$ for the $t = 1, 2, ..., T$ traces, and $\sum_{t=1}^{T} r_E(t) = 0$. Then the environmental noise induced bias in the moving average is bounded as $b \leq a_0 n_w^2/2$ where a_0 is the maximum of the derivative $|r'_E(t)|$. Let Δ_b^* denote the test statistic in Eq. (8) where the centering quantities $c^{(1)}$ and $c^{(2)}$ are each biased by the amount b. Then, (see Appendix B), $E(\Delta_b^*) = E(\Delta)$ and

$$\frac{Var(\Delta_b^*)}{Var(\Delta)} = 1 + \frac{b^2\eta^*}{Var(\Delta)} + o(n_w^4) \leq 1 + \frac{a_0^2 n_w^4 \eta^*}{4Var(\Delta)} + o(n_w^4), \qquad (11)$$

bounds the harm of using a too big n_w value, where η^* is

$$Var(L_A^{(1)}) + Var(L_A^{(2)}) + Var(L_B^{(1)}) + Var(L_B^{(2)}) + 2E(V_A^{(1)}V_A^{(2)}) + 2E(V_B^{(1)}V_B^{(2)}).$$

Matching the Eqs. (10) and (11), we can estimate the optimal window size from $n_w^5 \approx$

$$\frac{4[Var(V_A^{(1)})Var(V_A^{(2)}) + Var(V_B^{(1)})Var(V_B^{(2)}) + E^2(V_A^{(1)}V_A^{(2)}) + E^2(V_B^{(1)}V_B^{(2)})]}{a_0^2[Var(L_A^{(1)}) + Var(L_A^{(2)}) + Var(L_B^{(1)}) + Var(L_B^{(2)}) + 2E(V_A^{(1)}V_A^{(2)}) + 2E(V_B^{(1)}V_B^{(2)})]}.$$

As an example, we estimate this window size using parameters for data sets reported in literature. For simplicity, we assume that both leakage time points follow a similar power model, $V_A^{(i)} = \epsilon[HW_i - E(HW)]$, $i = 1, 2$, with HW_i as hamming weights related to masks and plaintexts as in the model of [7,18]. Hence $E(V_A^{(1)}V_A^{(2)}) = 0$ can be dropped, and $Var(V_A^{(1)}) = \epsilon^2 Var(HW) = 2\epsilon^2$ for the one-byte hamming weight. With the signal-noise-ratio ϵ/σ around 0.1 as in [7,9], the noise variance dominates so that $Var(L_A^{(1)}) \approx \sigma^2$. Since the two groups A and B follows the same power model, the optimal window size formula is simplified to $\{4[2(2\epsilon^2)^2]/[a_0^2 4\sigma^2]\}^{1/5} = [8(\epsilon/\sigma)^4\sigma^2/a_0^2]^{1/5} = [8(0.1)^4\sigma^2/a_0^2]^{1/5}$. For the 2373-th time point on the traces on the DPA V2 contest data shown in Fig. 1, the environmental fluctuation is approximately four periods of sinusoidal curve over one million time points with magnitude ≈ 100. So taking the maximum derivative of this curve, $a_0 \approx 1/400$. Fitting the power model at this time point gets $\sigma \approx 300$. Hence the optimal window size here is $[400^2 8(0.1)^4 300^2]^{1/5} \approx 30$ traces. This optimal window size does vary with the magnitude of the environmental fluctuation and the leakage signal-noise-ratio which are not known to a tester as a prior. But this example can serve as a rough benchmark, and a window size of a few dozens may be used in practice.

3.3   Computational Efficiency

The paired t-test also has computational advantages over Welch's t-test. As pointed out in [19], computational stability can become an issue when using raw moments for large measurement campaigns. The paired t-test computes mean $\bar{D}$ and variance s_D^2 of local differences D. In case there is no detectable leakage, L_A and L_B have the same mean. Hence, the differences D are mean-free[1]. Even computing $\bar{D} = \frac{1}{n_i}\sum d_i$ is thus numerically stable. The sample variance s_D^2 can be computed as $s_D^2 = \overline{D^2} - (\bar{D})^2$, where only the first term $\overline{D^2}$ is not mean-free. We used the incremental equation from [17, Eq. (1.3)] to avoid numerical problems. Moreover, by applying the incremental equation for $\bar{D}$ as well, we were able to exploit straightforward parallelism when computing $\bar{D}$ and variance s_D^2.

---

[1] If D is not mean-free, a strong leakage exists. Hence, a small number of observations suffices for leakage detection, making numerical problems irrelevant.

Table 1. Computation accuracy between our incremental method and two-pass algorithm

	1st order	2nd order	3rd order	4th order	5th order
Our method	50.0097	2.4679e+3	4.5981e+5	7.3616e+7	1.7974e+10
Two pass	50.0097	2.4679e+3	4.5981e+5	7.3616e+7	1.7974e+10

The situation essentially remains the same for higher order or multivariate analysis: The differences D are still mean-free in the no-leakage case. Through the use of local averages, the three-pass approach is not necessary, since moving averages are used instead of global averages (cf. Eq. (9)). Computing moving averages is a local operation, as only nearby traces are considered. When processing traces in large blocks of e.g. 10k traces, all data needed for local averages is within the same file and can easily be accessed when needed, making the algorithm essentially one-pass. Similarly as in [19], we also give the experimental results using our method on 100 million simulated traces with $\sim\mathcal{N}(100, 25)$. Specifically, we compute the second parameters s_D^2 using the difference leakages: $D = L_A - L_B$ for first order test while $D = [(L_A - \bar{L}_{A,n_w})^d - (L_B - \bar{L}_{B,n_w})^d]$ for d-th order tests with moving average of window size $n_w = 100$. Table 1 shows our method matches the two-pass algorithm which computes the mean first and then the variance of the preprocessed traces. Note that D is not normalized using the central moment CM_2 and thus the second parameter is significantly larger than that in [19]. In the experiments, the same numerical stability is achieved without an extra pass, by focusing on the difference leakages.

4   Experimental Verification

To show the advantages of the new approach, the performances of the paired t-test (3) and the unpaired t-test (1) on synthetic data are compared.

First, we generate data for first order leakage according to model (4), where the environmental noise r_E follows a sinusoidal pattern similar to Fig. 1. The sinusoidal period is set as $200,000$ traces, and the sinusoidal magnitude is set as the pure measurement noise standard deviation $\sigma_A = \sigma_B = 50$. Hamming weight (HW) leakage is assumed in model (4). The first group A uses a fixed plaintext input corresponds to $HW = 5$, while the second group B uses random plaintexts. The paired t-test (3) and the unpaired t-test (1) are applied to the first $n = 30000, 60000, ..., 300000$ pairs of traces. The experiment is repeated 1000 times, and the proportions of leakage detection (rejection by each t-test) are plotted in Fig. 2.

Without any environmental noise r_E, the paired and unpaired t-tests perform the same. Their success rate curves overlap each other. With the sinusoidal noise r_E, the unpaired t-test uses many more traces to detect the leakage, while the paired t-test does not suffer from such performance degradation.

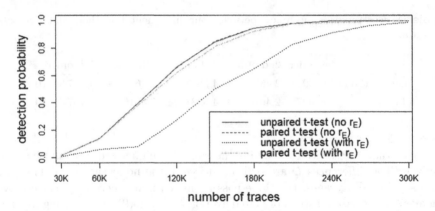

Fig. 2. T-test comparison for 1O leakage with and without a sinusoidal drift r_E. (Color figure online)

Notice that the environmental noise r_E often changes slowly as in Fig. 1. Hence, its effect is small for easy to detect leakage, when only a few hundreds or a few thousands of traces are needed. However, for hard to detect leakage, the effect has to be considered. We set a high noise level $\sigma_A = \sigma_B = 50$ to simulate a DUT with hard to detect first-order leakage. This allows the observable improvement by paired t-test over the unpaired t-test.

Second, we also generate data from the 2nd-order leakage model (7). The noise levels at the two leakage points, for both groups A and B, are set as $\sigma_1 = \sigma_2 = 10$ which are close to the levels in the physical implementation reported by [7]. We use the same sinusoidal environmental noise r_E as before. The first group A uses a fixed plaintext input corresponds to $HW = 1$, while the second group B uses random plaintexts. The proportions of leakage detection

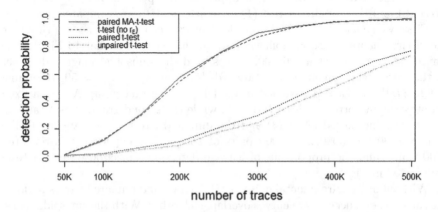

Fig. 3. T-test comparison for 2O leakage with a sinusoidal drift r_E. (Color figure online)

are plotted in Fig. 3. Again, we observe a serious degradation of t-test power to detect the leakage, when the environmental noise r_E is present. The paired t-test detects the leakage more often than the unpaired t-test in Fig. 3. However, the paired t-test also degrades comparing to the case without environmental noise r_E. That is due to the incorrect centering quantity for the 2O test as discussed in Sect. 3.2. Using the proposed method of centering at the moving average with window size 100, the paired MA-t-test has a performance close to the case where all environmental noise r_E is removed.

5   Practical Application

To show the advantage of the paired t-test in real measurement campaigns, we compare the performances of the unpaired and paired t-tests when analyzing an unprotected and an protected hardware implementation. The analysis focuses on the non-specific fixed vs. random t-test. We apply both tests to detect the first order leakage in the power traces acquired from an unprotected implementation of the NSA lightweight cipher Simon [2]. More specifically, a round-based implementation of Simon128/128 was used, which encrypts a 128-bit plaintext block with a 128-bit key in 68 rounds of operation. The second target is a masked engine of the same cipher. It is protected using three-share Threshold Implementation (TI) scheme, which is a round based variant of the TI Simon engine proposed in [20].

Both implementations are ported onto the SASEBO-GII board for power trace collection. The board is clocked at 3 MHz and a Tektronix oscilloscope samples the power consumption at 100 MS/s. Since Simon128/128 has 68 rounds, one power trace has about $68 \times \frac{1}{3\,\text{MHz}} \times 100\,\text{MS/s} \approx 2300$ time samples to cover the whole encryption and hence in the following experiments 2500 samples are taken in each measurement. The measurement setup is a modern setup that features a DC block and an amplifier. Note that the DC block will already take care of slow DC drifts that can affect the sensitivity of the unpaired t-test, as shown in Sect. 4. However, the DC block does not affect variations of the peak-to-peak height within traces, which are much more relevant for DPA. As the following experiments show, the paired t-test still shows improvement in such advanced setups.

5.1   Solving the Test-Order Bias

In [19], a random selection between fixed and random is proposed to avoid effects caused by states that occur in a fixed order, which we refer to as *test order*. For the paired (MA-)t-test, it is preferable to have a matching number of observations for both sets. We propose a fixed input sequence which is a repetition of $ABBA$ such that all the AB or BA pairs are constructed using neighboring inputs. For example in a sequence $ABBAABBA....ABBAABBA$, one alternately obtains AB and BA pairs with least variation. This ensures that all observations come in pairs and that the pairs are temporally close, so they share their environmental effects to a maximal possible degree. Moreover—even though the sequence is fixed and highly regular, the predecessor and successor for each measurement are perfectly balanced, corresponding to a 50 % probability of being either from the A or B set. This simpler setup removes the biases observed in [19] as efficiently as the random selection method. Experimental data of this section has been obtained using this scheme.

Note that the paired t-test can easily be applied in a random selection test order as well: After the trace collection, one can simply iteratively pair the leakages associated with the oldest fixed input and the oldest random input and then remove them from the sequence until no pairs can be constructed. An efficient way to do this is to separate all leakage traces into two subsets: $L_A = \{l_{A,1}, ...l_{A,n_A}\}$ and $L_B = \{l_{B,1}, ...l_{B,n_B}\}$ where $l_{A,i}$ and $l_{B,i}$ are the traces associated with i-th fixed input and i-th random input respectively in a chronological order and thus can be straightforwardly paired. Note that the cardinality of both sets are not always the same and hence only $n = min(n_A, n_B)$ AB pairs can be found. This approach is of less interest because time delay between fixed data and random data in a pair varies depending on the randomness of the input sequence.

5.2   First Order Analysis of an Unprotected Cipher

We first apply both paired and unpaired t-test to the unprotected engine which has strong first order leakage that can be exploited by DPA with only hundreds of traces. Usually the trace collection can be done quickly enough to avoid effects of environmental fluctuation in the measurements. However, to show the benefits of the paired t-test in this scenario, a hot air blower is used to heat up the crypto FPGA in SASEBO-GII board while the encryptions are executed. We designed two conditions to take the power measurements.

1. **Normal Lab Environment**, where measurements are performed in rapid succession, making the measurement campaign finish within seconds.
2. **Strong Environmental Fluctuation**, where a hot air blower was slowly moved towards and then away from the target FPGA to heat up and let it cool down again;

In each condition, 1000 measurements are taken alternately for the fixed plaintext and random plaintexts and later equally separated into two groups. In each group, the measurements are sorted in chronological order such that the j-th measurements of both groups are actually taken consecutively and share common variation. As explained in Sect. 5.1, the two measurements are a *matched-pair* and there are now 500 such pairs. Then both t-tests are applied to the first $n = 5, 6, 7, ..., 500$ pairs of measurements. For each n, the t-test returns a t-statistic vector of 2500 elements corresponding to 2500 time samples in the power traces because it is a univariate t-test. Our interest is the time sample that has the maximum t-statistic and thus the following results only focus on this specific time sample.

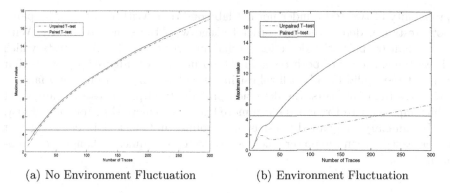

(a) No Environment Fluctuation   (b) Environment Fluctuation

Fig. 4. T-test comparison for 1O leakage on unprotected Simon for a single measurement campaign of up to 300 pairs of traces. The paired t-test performs as well or better in both scenarios. However, the paired t-test is more robust to environmental noise. (Color figure online)

Figure 4 shows the t-statistics at the strongest leakage point as n increases. In Fig. 4(a) where there is no environmental fluctuation, both unpaired and paired t-test have the same performance as the t-statistic curves almost overlap. However, in Fig. 4(b) where the varying temperature changed the power traces greatly, the paired t-test (blue solid line) shows robustness and requires less traces to exceed the threshold of 4.5 while the performance of the unpaired t-test is greatly reduced in the sense that more traces are needed to go beyond the threshold. Figure 5 shows the detection probability of the t-tests in the same scenario. First, 1000 repetitions of the above experiment are performed and the number of experiments that result in a t-statistic above 4.5 is counted. Detection probability equals this number divided by 1000. Figure 5(a) shows the detection

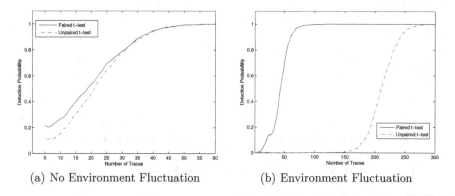

(a) No Environment Fluctuation   (b) Environment Fluctuation

Fig. 5. T-test detection probability for 10 leakage. Again, the paired t-test performs at least as well as the unpaired, while being much more robust in the presence of environmental noise.

probability of two tests under normal lab condition. With more than 30 pairs, both tests can detect the first order leakage with the same probability. With more than 60 pairs the detection probability rises to 1 for both tests which shows the efficiency of both tests on the normal traces. Figure 5(b) shows that paired t-test (solid line) is still robust in spite of varying environmental factors. With less than 100 pairs, the detection probability of paired t-test is already 1 while unpaired t-test requires much more traces to achieve the same probability.

In summary, the paired t-test is more robust and efficient in detecting first order leakage when the power traces are collected in a quickly changing environment.

5.3 Second Order Analysis on a First-Order Resistant Design

In order to validate the effectiveness of the paired t-test in a longer measurement campaign, where environmental fluctuations are very likely to occur, a first-order-leakage-resistant Simon engine protected by a three-share Threshold Implementation scheme is used as the target. Five million power traces are collected in a room without windows and without expected fluctuations in temperature over a period 5 h. As before, one measurement campaign is performed in a stable lab environment where the environmental conditions are kept as stable as possible. In the other scenario, we again used the hot air blower in intervals of several minutes to simulate stronger environmental noise. This is because the environmental noise might not be strong during the 5-h collection period. However, in scenarios where hundreds of millions of measurements are needed and taken over a period of several days, then environmental fluctuation can be found, as in Fig. 1.

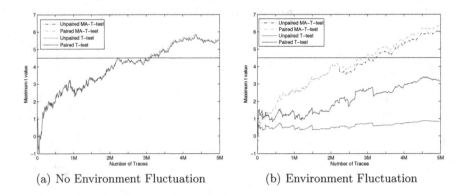

(a) No Environment Fluctuation   (b) Environment Fluctuation

Fig. 6. T-test detection probability for 2O leakage (Color figure online)

As before, the 5 million traces are equally divided into two groups for fixed and random plaintext respectively. The first order t-test does not indicate any leakage ($|t| < 3$), as expected. Figure 6 shows the t-statistics of the second order t-tests as the number of traces increases in both the stable lab environment and the simulated lab environment noise scenario. In the first experiment in a stable environment, depicted in Fig. 6(a), we compare both tests using global average and moving average. The curve of four tests almost overlap and they perform about the same with about three million traces needed to achieve a t-statistic above 4.5. This shows that paired t-test works as well as unpaired one for constant collection environment. Also, the moving average based tests perform very similar to the global average based tests, with a minor improvement in the relevant many-traces case. Figure 6(b) depicts the results for the experiment with strong environmental fluctuations. The paired MA-t-test performs best and goes beyond 4.5 faster than the unpaired one. The other two tests using global average are still below the threshold with 5 million traces. The paired t-test still clearly outperforms the unpaired t-test. In sum, the paired t-test based on moving average is the most robust to fluctuation and significantly improves the performance of higher order analysis.

6   Conclusion

Welch's t-test has recently received a lot of attention as standard side channel security evaluation tool. In this work we showed that noise resulting from environmental fluctuations can negatively impact the performance of Welch's t-test. The resulting increased number of observations to detect a leakage are inconvenient and can, in the worst case, result in false conclusions about a device's resistance. We proposed a paired t-test to improve the standard methodology for leakage detection. The resulting matched-pairs design removes the environmental noise effect in leakage detection. Furthermore, we showed that moving averages increase the robustness against environmental noise for higher order or multivariate analysis, while not showing any negative impact in the absence of noise. The improvement is shown through mathematical analysis, simulation, and on practical power measurements: both paired and unpaired t-test with and without the moving averages approach are compared for first order and second order analysis. Our results show that the proposed (moving average based) paired t-test performed as well or better in all analyzed scenarios. The new method does not increase computational complexity and is numerically more stable than Welch's t-test. Since our method is more robust to environmental noise and can detect leakage faster than unpaired test in the presence of noise, we propose the replacement of Welch's t-test with the moving average based paired t-test as a standard leakage detection tool.

Acknowledgments. This work is supported by the National Science Foundation under grant CNS-1314655, CNS-1314770 and CNS-1261399.

Appendix

A   Proof of Theorem 1

We are comparing the leakage detection statistic (9)

$$D = (L_A^{(1)} - \bar{L}_A^{(1)})(L_A^{(2)} - \bar{L}_A^{(2)}) - (L_B^{(1)} - \bar{L}_B^{(1)})(L_B^{(2)} - \bar{L}_B^{(2)}),$$

with the theoretical optimal leakage detection statistic Δ in Eq. (8).

Without loss of generality, let $c^{(1)} = c^{(2)} = 0$ in model (6), since these constants are cancelled in each of the differences $(L_A^{(j)} - \bar{L}_A^{(j)})$ and $(L_B^{(j)} - \bar{L}_B^{(j)})$ for $j = 1, 2$. Then (8) is simplified as $\Delta = L_A^{(1)} L_A^{(2)} - L_B^{(1)} L_B^{(2)}$. Hence

$$E(\Delta) = E(L_A^{(1)} L_A^{(2)}) - E(L_B^{(1)} L_B^{(2)})$$
$$Var(\Delta) = Var(L_A^{(1)} L_A^{(2)}) + Var(L_B^{(1)} L_B^{(2)}). \tag{12}$$

We first reexpress $(L_A^{(1)} - \bar{L}_A^{(1)})$ as the difference between two independent terms. We denote $\tilde{L}_A^{(1)} = \frac{1}{n_w - 1} \sum_{i=1}^{n_w - 1} L_{A,i}^{(1)}$ as the average of $n_w - 1$ traces excluding the original trace, where $L_{A,i}^{(1)}$ ($i = 1, ..., n_w - 1$) are independent random variables coming from the same distribution as $L_A^{(1)}$. Since $\bar{L}_A^{(1)}$ is the average over n_w nearby traces including the original trace, $\bar{L}_A^{(1)} = \frac{1}{n_w}[L_A^{(1)} + \sum_{i=1}^{n_w - 1} L_{A,i}^{(1)}] = \frac{n_w - 1}{n_w}(L_A^{(1)} - \tilde{L}_A^{(1)})$, with $\tilde{L}_A^{(1)}$ independent of $L_A^{(1)}$. $E(\tilde{L}_A^{(1)}) = E(L_A^{(1)})$ and $Var(\tilde{L}_A^{(1)}) = \frac{1}{n_w - 1} Var(L_A^{(1)})$. Similarly, $\tilde{L}_A^{(2)}$, $\tilde{L}_B^{(1)}$ and $\tilde{L}_B^{(2)}$ denotes the average of corresponding quantities over the $n_w - 1$ traces excluding the original trace. The we can rewrite the leakage detection statistic in (9) as

$$D = (\frac{n_w - 1}{n_w})^2 [(L_A^{(1)} - \tilde{L}_A^{(1)})(L_A^{(2)} - \tilde{L}_A^{(2)}) - (L_B^{(1)} - \tilde{L}_B^{(1)})(L_B^{(2)} - \tilde{L}_B^{(2)})]. \tag{13}$$

Therefore as $n_w \to \infty$, $D \to \Delta$.

Next, we show that $E(D)$ and $Var(D)$ differ from their limits $E(\Delta)$ and $Var(\Delta)$ by a factor of $O(1/n_w)$ only. Let $D^* = \frac{n_w}{n_w - 1} D$. Then we have

$$E(D^*) = E(\Delta), \tag{14}$$

$$Var(D^*) - Var(\Delta)$$
$$= \frac{2}{n_w}[Var(V_A^{(1)}) Var(V_A^{(2)}) + Var(V_B^{(1)}) Var(V_B^{(2)}) + E^2(V_A^{(1)} V_A^{(2)})$$
$$+ E^2(V_B^{(1)} V_B^{(2)}) - Var(V_A^{(1)} V_A^{(2)}) - Var(V_B^{(1)} V_B^{(2)})] + O(\frac{1}{n_w^2}). \tag{15}$$

The proofs of these two equations are provided in the next two subsections.

Combining Eqs. (12), (14) and (15), we arrived at Eq. (10) and Theorem 1 is proved.

A.1 Proof of Eq. (14) on Mean of D^*

We now calculate the first term in $E(D)$.

$$E(\tilde{L}_A^{(1)}\tilde{L}_A^{(2)}) = (\frac{1}{n_w - 1})^2 \sum_{i=1}^{n_w-1} \sum_{j=1}^{n_w-1} E(L_{A,i}^{(1)}L_{A,j}^{(2)}).$$

For $i \neq j$, $L_{A,i}^{(1)}$ is independence of $L_{A,j}^{(2)}$ so that $E(L_{A,i}^{(1)}L_{A,j}^{(2)}) = E(L_{A,i}^{(1)})E(L_{A,j}^{(2)}) = (0)(0) = 0$ and drops from the summation. Hence

$$E(\tilde{L}_A^{(1)}\tilde{L}_A^{(2)}) = (\frac{1}{n_w - 1})^2 \sum_{i=1}^{n_w-1} E(L_{A,i}^{(1)}L_{A,i}^{(2)}) = \frac{1}{n_w - 1}E(L_A^{(1)}L_A^{(2)}). \quad (16)$$

Also, since $\tilde{L}_A^{(1)}$ is independent of $L_A^{(2)}$, $E(\tilde{L}_A^{(1)}L_A^{(2)}) = E(\tilde{L}_A^{(1)})E(L_A^{(2)}) = 0$. Similarly $E(L_A^{(1)}\tilde{L}_A^{(2)}) = 0$. Therefore,

$$\begin{aligned}
E[(L_A^{(1)} - \tilde{L}_A^{(1)})(L_A^{(2)} - \tilde{L}_A^{(2)})] &= E(L_A^{(1)}L_A^{(2)}) - 0 - 0 + E(\tilde{L}_A^{(1)}\tilde{L}_A^{(2)}) \\
&= E(L_A^{(1)}L_A^{(2)}) + \frac{1}{n_w - 1}E(L_A^{(1)}L_A^{(2)}) \\
&= \frac{n_w}{n_w - 1}E(L_A^{(1)}L_A^{(2)}).
\end{aligned}$$

Similarly, $E[(L_B^{(1)} - \tilde{L}_B^{(1)})(L_B^{(2)} - \tilde{L}_B^{(2)})] = \frac{n_w}{n_w-1}E(L_B^{(1)}L_B^{(2)})$. Combine these two expressions with Eq. (13) and $D^* = \frac{n_w}{n_w-1}D$, we get Eq. (14)

$$E(D^*) = (\frac{n_w - 1}{n_w})\frac{n_w}{n_w - 1}E[L_A^{(1)}L_A^{(2)} - L_B^{(1)}L_B^{(2)}] = E(\Delta).$$

A.2 Proof of Eq. (15) on Variance of D^*

$$Var(D^*) = (\frac{n_w - 1}{n_w})^2\{Var[(L_A^{(1)} - \tilde{L}_A^{(1)})(L_A^{(2)} - \tilde{L}_A^{(2)})] + Var[(L_B^{(1)} - \tilde{L}_B^{(1)})(L_B^{(2)} - \tilde{L}_B^{(2)})]\}. \quad (17)$$

For the first term, the variance of the sum $L_A^{(1)}L_A^{(2)} - \tilde{L}_A^{(1)}L_A^{(2)} - L_A^{(1)}\tilde{L}_A^{(2)} + L_A^{(1)}L_A^{(2)}$ is the covariance of the sum with itself. For the four terms in $L_A^{(1)}L_A^{(2)} - \tilde{L}_A^{(1)}L_A^{(2)} - L_A^{(1)}\tilde{L}_A^{(2)} + L_A^{(1)}L_A^{(2)}$, the covariance for most pairs of different terms are zero. For example,

$$\begin{aligned}
Cov(L_A^{(1)}L_A^{(2)}, \tilde{L}_A^{(1)}L_A^{(2)}) &= E(L_A^{(1)}L_A^{(2)}\tilde{L}_A^{(1)}L_A^{(2)}) - E(L_A^{(1)}L_A^{(2)})E(\tilde{L}_A^{(1)}L_A^{(2)}) \\
&= E(L_A^{(1)}L_A^{(2)}L_A^{(2)})0 - E(L_A^{(1)}L_A^{(2)})E(L_A^{(2)})0 = 0.
\end{aligned}$$

and $Cov(L_A^{(1)}L_A^{(2)}, \tilde{L}_A^{(1)}\tilde{L}_A^{(2)}) = 0$ due to the independence between $L_A^{(1)}L_A^{(2)}$ and $\tilde{L}_A^{(1)}\tilde{L}_A^{(2)}$. The only non-zero cross-term covariance is

$$Cov(\tilde{L}_A^{(1)}L_A^{(2)}, L_A^{(1)}\tilde{L}_A^{(2)}) = E(\tilde{L}_A^{(1)}L_A^{(2)}L_A^{(1)}\tilde{L}_A^{(2)}) - 0 = E(L_A^{(1)}L_A^{(2)})E(\tilde{L}_A^{(1)}\tilde{L}_A^{(2)})$$
$$= \frac{1}{n_w - 1}E^2(L_A^{(1)}L_A^{(2)}),$$

with the last step coming from Eq. (16). Therefore,

$$Var[(L_A^{(1)} - \tilde{L}_A^{(1)})(L_A^{(2)} - \tilde{L}_A^{(2)})]$$
$$= Var(L_A^{(1)}L_A^{(2)}) + Var(\tilde{L}_A^{(1)}L_A^{(2)}) + Var(L_A^{(1)}\tilde{L}_A^{(2)}) + Var(\tilde{L}_A^{(1)}\tilde{L}_A^{(2)})$$
$$+ \frac{2}{n_w - 1}E^2(L_A^{(1)}L_A^{(2)})$$

By independence, $Var(\tilde{L}_A^{(1)}L_A^{(2)}) = Var(\tilde{L}_A^{(1)})Var(L_A^{(2)}) = \frac{1}{n_w-1}Var(L_A^{(1)})$
$Var(L_A^{(2)})$, and $Var(L_A^{(1)}\tilde{L}_A^{(2)}) = \frac{1}{n_w-1}Var(L_A^{(1)})Var(L_A^{(2)})$.

For $Var(\tilde{L}_A^{(1)}\tilde{L}_A^{(2)})$, note that

$$\tilde{L}_A^{(1)}\tilde{L}_A^{(2)} = (\frac{1}{n_w - 1})^2 \sum_{i=1}^{n_w-1} \sum_{j=1}^{n_w-1} L_{A,i}^{(1)}L_{A,j}^{(2)}.$$

The covariance between any two different terms in the sum is zero. Hence

$$Var(\tilde{L}_A^{(1)}\tilde{L}_A^{(2)}) = (\frac{1}{n_w - 1})^4 [\sum_i Var(L_{A,i}^{(1)}L_{A,i}^{(2)}) + \sum_{i \neq j} Var(L_{A,i}^{(1)}L_{A,j}^{(2)})]$$
$$= \frac{1}{(n_w - 1)^3}Var(L_A^{(1)}L_A^{(2)}) + \frac{n_w - 2}{(n_w - 1)^3}Var(L_A^{(1)})Var(L_A^{(2)}).$$

Combine together, we have

$$Var[(L_A^{(1)} - \tilde{L}_A^{(1)})(L_A^{(2)} - \tilde{L}_A^{(2)})]$$
$$= Var(L_A^{(1)}L_A^{(2)}) + \frac{2}{n_w - 1}Var(L_A^{(1)})Var(L_A^{(2)}) + \frac{2}{n_w - 1}E^2(L_A^{(1)}L_A^{(2)})$$
$$+ \frac{n_w - 2}{(n_w - 1)^3}Var(L_A^{(1)})Var(L_A^{(2)}) + \frac{1}{(n_w - 1)^3}Var(L_A^{(1)}L_A^{(2)})$$
$$= Var(L_A^{(1)}L_A^{(2)}) + \frac{2}{n_w}Var(L_A^{(1)})Var(L_A^{(2)}) + \frac{2}{n_w}E^2(L_A^{(1)}L_A^{(2)}) + O(\frac{1}{n_w^2})$$

Hence the first term in $Var(D^*)$ becomes

$$(\frac{n_w - 1}{n_w})^2 Var[(L_A^{(1)} - \tilde{L}_A^{(1)})(L_A^{(2)} - \tilde{L}_A^{(2)})]$$
$$= (\frac{n_w - 1}{n_w})^2 Var(L_A^{(1)}L_A^{(2)}) + \frac{2}{n_w}Var(L_A^{(1)})Var(L_A^{(2)}) + \frac{2}{n_w}E^2(L_A^{(1)}L_A^{(2)}) + O(\frac{1}{n_w^2})$$
$$= Var(L_A^{(1)}L_A^{(2)}) + \frac{2}{n_w}[Var(L_A^{(1)})Var(L_A^{(2)}) + E^2(L_A^{(1)}L_A^{(2)}) - Var(L_A^{(1)}L_A^{(2)})] + O(\frac{1}{n_w^2}).$$
$$\tag{18}$$

For further simplification, let σ_1^2 and σ_2^2 denote the variances of noises $r^{(1)}$ and $r^{(2)}$ in the second-order leakage model (6). Then $Var(L_A^{(1)}) = \sigma_1^2 + Var(V^{(1)})$, $Var(L_A^{(2)}) = \sigma_2^2 + Var(V^{(2)})$, $E(L_A^{(1)}L_A^{(2)}) = E(V^{(1)}V^{(2)})$,

$$E[(L_A^{(1)}L_A^{(2)})^2] = E[(V_A^{(1)} + r_A^{(1)})^2(V_A^{(2)} + r_A^{(2)})^2]$$
$$= E[(V_A^{(1)})^2(V_A^{(2)})^2 + (r_A^{(1)})^2(V_A^{(2)})^2 + (V_A^{(1)})^2(r_A^{(2)})^2 + (r_A^{(1)})^2(r_A^{(2)})^2] + 0$$
$$= E[(V_A^{(1)})^2(V_A^{(2)})^2] + \sigma_1^2 Var(V_A^{(2)}) + \sigma_2^2 Var(V_A^{(1)}) + \sigma_1^2\sigma_2^2.$$

Hence

$$Var[L_A^{(1)}L_A^{(2)}] = Var(V_A^{(1)}V_A^{(2)}) + \sigma_1^2 Var(V_A^{(2)}) + \sigma_2^2 Var(V_A^{(1)}) + \sigma_1^2\sigma_2^2.$$

Combine the above five expressions,

$$Var(L_A^{(1)})Var(L_A^{(2)}) + E^2(L_A^{(1)}L_A^{(2)}) - Var(L_A^{(1)}L_A^{(2)})$$
$$= Var(V^{(1)})Var(V^{(2)}) + E(V^{(1)}V^{(2)}) - Var(V_A^{(1)}V_A^{(2)})$$

Combine this with (17) and (18) we have Eq. (15),

$$Var(D^*) - [Var(L_A^{(1)}L_A^{(2)}) + Var(L_B^{(1)}L_B^{(2)})]$$
$$= \frac{2}{n_w}[Var(V_A^{(1)})Var(V_A^{(2)}) + E^2(V_A^{(1)}V_A^{(2)}) - Var(V_A^{(1)}V_A^{(2)})$$
$$+ Var(V_B^{(1)})Var(V_B^{(2)}) + E^2(V_B^{(1)}V_B^{(2)}) - Var(V_B^{(1)}V_B^{(2)})] + O(\frac{1}{n_w^2}).$$

B   Derivation of Eq. (11)

As in the previous section, we let $c^{(1)} = c^{(2)} = 0$ without loss of generality, so that $E(L_A^{(1)}) = E(L_A^{(2)}) = 0$. Then

$$E[(L_A^{(1)} - b)(L_A^{(2)} - b)] = E(L_A^{(1)}L_A^{(2)}) - bE(L_A^{(1)}) - bE(L_A^{(2)}) + b^2 = E(L_A^{(1)}L_A^{(2)}) + b^2$$
$$= E(L_A^{(1)}L_A^{(2)}) + b^2.$$

Hence

$$E(\Delta_b^*) = E[(L_A^{(1)} - b)(L_A^{(2)} - b)] - E[(L_B^{(1)} - b)(L_B^{(2)} - b)]$$
$$= E(L_A^{(1)}L_A^{(2)}) + b^2 - E(L_B^{(1)}L_B^{(2)}) - b^2$$
$$= E(L_A^{(1)}L_A^{(2)}) - E(L_B^{(1)}L_B^{(2)}) = E(\Delta). \tag{19}$$

Next,

$$Var[(L_A^{(1)} - b)(L_A^{(2)} - b)]$$
$$=E[(L_A^{(1)} - b)^2(L_A^{(2)} - b)^2] - [E(L_A^{(1)}L_A^{(2)}) + b^2]^2$$
$$=E[((L_A^{(1)})^2 - 2bL_A^{(1)} + b^2)((L_A^{(2)})^2 - 2bL_A^{(2)} + b^2)] - E[(L_A^{(1)}L_A^{(2)})^2] - 2bE(L_A^{(1)}L_A^{(2)}) - b^4$$
$$=Var(L_A^{(1)}L_A^{(2)}) - 2bE[L_A^{(1)}L_A^{(2)}(L_A^{(1)} + L_A^{(2)})] + b^2E[(L_A^{(1)})^2 + (L_A^{(2)})^2 + 2L_A^{(1)}L_A^{(2)}]$$
$$=Var(L_A^{(1)}L_A^{(2)}) + b^2[Var(L_A^{(1)}) + Var(L_A^{(2)}) + 2E(L_A^{(1)}L_A^{(2)})] + O(b).$$

Hence we get the variance

$$Var(\Delta_b^*) = Var(\Delta) + b^2[Var(L_A^{(1)}) + Var(L_A^{(2)}) + 2E(L_A^{(1)}L_A^{(2)})$$
$$+ Var(L_B^{(1)}) + Var(L_B^{(2)}) + 2E(L_B^{(1)}L_B^{(2)})] + O(b). \qquad (20)$$

References

1. Balasch, J., Gierlichs, B., Grosso, V., Reparaz, O., Standaert, F.-X.: On the cost of lazy engineering for masked software implementations. In: Joye, M., Moradi, A. (eds.) CARDIS 2014. LNCS, vol. 8968, pp. 64–81. Springer, Heidelberg (2015). http://dx.doi.org/10.1007/978-3-319-16763-3_5
2. Beaulieu, R., Shors, D., Smith, J., Treatman-Clark, S., Weeks, B., Wingers, L.: The Simon and Speck families of lightweight block ciphers. IACR Cryptol. ePrint Arch. **2013**, 404 (2013)
3. Bilgin, B., Gierlichs, B., Nikova, S., Nikov, V., Rijmen, V.: A more efficient AES threshold implementation. In: Pointcheval, D., Vergnaud, D. (eds.) AFRICACRYPT. LNCS, vol. 8469, pp. 267–284. Springer, Heidelberg (2014)
4. Bilgin, B., Gierlichs, B., Nikova, S., Nikov, V., Rijmen, V.: Higher-order threshold implementations. In: Sarkar, P., Iwata, T. (eds.) ASIACRYPT 2014, Part II. LNCS, vol. 8874, pp. 326–343. Springer, Heidelberg (2014)
5. Chen, C., Eisenbarth, T., von Maurich, I., Steinwandt, R.: Masking large keys in hardware: a masked implementation of McEliece. In: Dunkelman, O., et al. (eds.) SAC 2015. LNCS, vol. 9566, pp. 293–309. Springer, Heidelberg (2016). doi:10.1007/978-3-319-31301-6_18
6. Cooper, J., DeMulder, E., Goodwill, G., Jaffe, J., Kenworthy, G., Rohatgi, P.: Test Vector Leakage Assessment (TVLA) methodology in practice. In: International Cryptographic Module Conference (2013). http://icmc-2013.org/wp/wp-content/uploads/2013/09/goodwillkenworthtestvector.pdf
7. Ding, A.A., Zhang, L., Fei, Y., Luo, P.: A statistical model for higher order DPA on masked devices. In: Batina, L., Robshaw, M. (eds.) CHES 2014. LNCS, vol. 8731, pp. 147–169. Springer, Heidelberg (2014). http://dx.doi.org/10.1007/978-3-662-44709-3_9
8. Durvaux, F., Standaert, F.-X.: From improved leakage detection to the detection of points of interests in leakage traces. In: Fischlin, M., Coron, J.-S. (eds.) EUROCRYPT 2016. LNCS, vol. 9665, pp. 240–262. Springer, Heidelberg (2016). doi:10.1007/978-3-662-49890-3_10
9. Fei, Y., Ding, A.A., Lao, J., Zhang, L.: A statistics-based success rate model for DPA and CPA. J. Crypt. Eng. **5**(4), 227–243 (2015). doi:10.1007/s13389-015-0107-0

10. Goodwill, G., Jun, B., Jaffe, J., Rohatgi, P.: A testing methodology for side-channel resistance validation. In: NIST Non-Invasive Attack Testing Workshop, September 2011. http://csrc.nist.gov/news_events/non-invasive-attack-testing-workshop/papers/08_Goodwill.pdf
11. Heuser, A., Kasper, M., Schindler, W., Stöttinger, M.: A new difference method for side-channel analysis with high-dimensional leakage models. In: Dunkelman, O. (ed.) CT-RSA 2012. LNCS, vol. 7178, pp. 365–382. Springer, Heidelberg (2012). http://dx.doi.org/10.1007/978-3-642-27954-6_23
12. Kutner, M.H., Nachtsheim, C.J., Neter, J., Li, W.: Applied Linear Statistical Models. McGraw-Hill/Irwin, New York (2005)
13. Leiserson, A.J., Marson, M.E., Wachs, M.A.: Gate-level masking under a path-based leakage metric. In: Batina, L., Robshaw, M. (eds.) CHES 2014. LNCS, vol. 8731, pp. 580–597. Springer, Heidelberg (2014)
14. Mather, L., Oswald, E., Bandenburg, J., Wójcik, M.: Does my device leak information? an a priori statistical power analysis of leakage detection tests. In: Sako, K., Sarkar, P. (eds.) ASIACRYPT 2013, Part I. LNCS, vol. 8269, pp. 486–505. Springer, Heidelberg (2013). http://dx.doi.org/10.1007/978-3-642-42033-7_25
15. Moradi, A., Hinterwälder, G.: Side-channel security analysis of ultra-low-power FRAM-based MCUs. In: Mangard, S., Poschmann, A.Y. (eds.) COSADE 2015. LNCS, vol. 9064, pp. 239–254. Springer, Heidelberg (2015). http://dx.doi.org/10.1007/978-3-319-21476-4_16
16. Nascimento, E., Lopez, J., Dahab, R.: Efficient and secure elliptic curve cryptography for 8-bit AVR microcontrollers. In: Chakraborty, R.S., et al. (eds.) SPACE 2015. LNCS, vol. 9354. Springer, Heidelberg (2015). http://dx.doi.org/10.1007/978-3-319-24126-5_17
17. Pébay, P.: Formulas for robust, one-pass parallel computation of covariances and arbitrary-order statistical moments. Sandia report SAND2008-6212, Sandia National Laboratories (2008)
18. Prouff, E., Rivain, M., Bevan, R.: Statistical analysis of second order differential power analysis. IEEE Trans. Comput. 58(6), 799–811 (2009)
19. Schneider, T., Moradi, A.: Leakage assessment methodology. In: Güneysu, T., Handschuh, H. (eds.) CHES 2015. LNCS, vol. 9293, pp. 495–513. Springer, Heidelberg (2015). http://dblp.uni-trier.de/db/conf/ches/ches2015.html SchneiderM15
20. Shahverdi, A., Taha, M., Eisenbarth, T.: Silent Simon: threshold implementation under 100 slices. In: 2015 IEEE International Symposium on Hardware Oriented Security and Trust (HOST), pp. 1–6, May 2015

Design and Implementation
of a Waveform-Matching
Based Triggering System

Arthur Beckers[✉], Josep Balasch, Benedikt Gierlichs, and Ingrid Verbauwhede

Department of Electrical Engineering-ESAT/COSIC and iMinds, KU Leuven,
Kasteelpark Arenberg 10, 3001 Heverlee, Leuven, Belgium
{arthur.beckers,josep.balasch,
benedikt.gierlichs,ingrid.verbauwhede}@esat.kuleuven.be

Abstract. Implementation attacks such as side channel attacks and fault attacks require triggering mechanisms to activate the acquisition device or fault injection equipment. Most academic works work with a very simple and reliable trigger mechanism where the device under test itself provides a dedicated signal. This however is not possible in real attack scenarios. Here the alternative is to use IO signals or coarse features of the side channel signal (co-processor switches on, power consumption goes up) for triggering. However, fault injection in particular requires very accurate timing. Our work deals with the many scenarios where such simple triggering mechanisms are not available or not effective. We present our design, architecture and FPGA implementation of a waveform-matching based triggering system. Our configurable trigger box is able to sample and match an arbitrary waveform with a latency of 128 ns. We provide results of our experimental evaluation on devices and side channel signals of different nature, and discuss the influence of several parameters.

Keywords: Triggering · Waveform matching · Fault injection

1 Introduction

Implementation attacks are well-known techniques that can pose a serious threat to the security of embedded devices. Side channel attacks rely on the analysis of physical observations of the device during cryptographic executions, for instance, running time [15], power consumption [16] or electromagnetic emanations [14,17]. Fault attacks [9] on the other hand rely on injecting faults during cryptographic computations. Examples are clock glitches [7,8], voltage spikes [19], electromagnetic pulses [13,18] or optical attacks [5,20].

To perform implementation attacks an adversary requires some sort of *triggering mechanism* capable of activating the side channel acquisition device or the equipment for fault injection. Depending on the concrete attack scenario, precise timing may be essential. The prevalent approach in academic works is to

© Springer International Publishing Switzerland 2016
F.-X. Standaert and E. Oswald (Eds.): COSADE 2016, LNCS 9689, pp. 184–198, 2016.
DOI: 10.1007/978-3-319-43283-0_11

generate a trigger signal from within the device under test. This allows to concentrate on the evaluation of a certain attack or countermeasure, while abstracting from practical issues. This approach is however not possible in realistic scenarios. Here, one would typically use the built-in triggering functionalities of an oscilloscope to detect simple features such as logic events in the IO line, well-defined shapes in the side channel signals (sudden amplitude changes, gaps of certain width, etc.), or a combination thereof. If the selected event does not occur just before or after the targeted operation, hold-off timers can be employed to shift the trigger closer to the time of interest.

While this approach may achieve high accuracy and reproducibility in certain settings, it suffers from two main limitations. First, the range of trigger options is limited to the capabilities of the oscilloscope, e.g. mostly edge, pulse and logic triggering. This may not be sufficient to trigger on devices where existing signals lack coarse features. And second, the insertion of hold-off timers implicitly assumes deterministic program executions. Devices with non-deterministic behaviour (due to preemptive multitasking, caches, branch predictors, etc.) or implementations with built-in countermeasures (random delays [10,11], clock jitter, etc.) can easily make triggering a practical bottleneck.

A more suitable alternative consists in using a *pattern-based* triggering mechanism which can detect arbitrary waveforms in the side channel signals. The method runs in two stages. First, one selects a suitable *reference* or *pattern* from a window of interest in the side channel signal. And second, one employs a waveform-matching algorithm to detect such pattern on an incoming side channel signal. In the following we denote these stages as *capture mode* and *matching mode*. It is important that the analog-to-digital conversion process in both modes is the same. The selection of the underlying waveform-matching algorithm depends on the use case requirements. Fast response time is particularly desirable for fault injection attacks. Flexibility allows the method to adapt to different setups. Additionally, robustness is required to compensate for the noise that is inherently present in the signals.

To the best of our knowledge the only publicly documented solution for pattern-based triggering is icWaves [4], developed and commercialized by Riscure. This solution implements a waveform-matching algorithm based on the sum of absolute differences. It can detect pattern(s) up to 1×512 samples or 2×256 samples long and has a response time of around 500 ns. The usage of this device for laser fault injection attacks has been documented by van Woudenberg et al. in [21].

Our Contribution. In this work we put forward a waveform-matching based triggering system for use in the context of implementation attacks. Our solution is specifically designed for low latency, i.e. to minimize the response time once the pattern occurs in the side channel, and it is based on an interval matching algorithm. We provide a detailed description of our design and architecture choices, as well as the implementation of a functional trigger box on an FPGA development board. Our solution supports detection of arbitrary waveforms, and incorporates multiple options to ensure flexibility and ease of adaptation to different scenarios. We illustrate these aspects by performing an empirical evaluation on two different cryptographic devices (dedicated Java Card smart card, high-speed general purpose ARM processor) with side channel signals of different nature (power measurements using shunt resistor, contactless power measurements using EM probe). Finally, we discuss the influence of several parameters on the triggering behaviour.

2   Waveform Matching

The essence of waveform matching is to compare a fixed *reference* (or *pattern*) signal g of length N samples with a continuous incoming signal h. There exist many different algorithms for pattern matching in the literature, but the vast majority work according to the same basic principle. The algorithm calculates a measure of correspondence between g and (a part of) h and represents it by a single score $T(k)$, where k represents a time shift from the starting execution point. The comparison of $T(k)$ with a pre-defined threshold determines whether the signals are considered a match. In our case, we work with discrete-time signals resulting from an analog-to-digital conversion. Therefore the threshold selection needs to account for the effect of quantization noise as well as noise caused by system and environmental variations.

In the following we review different options for waveform matching and discuss their suitability to our use case. We concentrate on algorithms that allow for low-latency matching and can be efficiently implemented in hardware. All considered algorithms perform a sample-wise comparison of the reference with the incoming signal, i.e. the score at a certain time shift k can be computed as:

$$T(k) = \sum_{m=1}^{N} score_m, \qquad \text{where } score_m = f\big(g(m), h(m+k)\big).$$

Cross-Correlation. This statistical function is perhaps the most natural algorithm to measure the similarity of two series. It uses the product of two samples to compute a measure of their resemblance as:

$$score_m = g(m)h(m+k). \tag{1}$$

Sign Comparison (see Fig. 1a). This method transforms the reference signal g into a binary sequence g' by assigning $g'(m) = 1$ if $g(m) > \mu$, and $g'(m) = 0$ otherwise. Here μ is the mean value of g. The incoming signal h is transformed into h' in the same manner. The sample score is calculated using g' and h' as:

$$score_m = \begin{cases} 1 & \text{if } g'(m) == h'(m+k) \\ 0 & \text{otherwise}. \end{cases} \tag{2}$$

Sum of Absolute Differences (SAD) (see Fig. 1b). This algorithm performs a sample-wise subtraction between reference and incoming signals, taking as a score the absolute value of the difference as:

$$score_m = \big|g(m) - h(m + k)\big|. \tag{3}$$

Interval Matching (see Fig. 1c). This algorithm defines an interval with a chosen *offset* above and below the reference. The score is calculated by checking whether the sample of the incoming signal lies within a valid interval as:

$$score_m = \begin{cases} 1 & \text{if } g(m) + offset \geq h(m + k) \geq g(m) - offset \\ 0 & \text{otherwise}. \end{cases} \tag{4}$$

Algorithm Selection. Our primary selection criteria for the pattern matching algorithm is low latency, but we also consider aspects such as flexibility, implementability and suitability to our application. Despite its robustness, we discard cross-correlation due to its use of multiplications. Note that the maximum pattern length is determined by the availability of multipliers on the implementation platform, which can be quite low. Additionally, the final score $T(k)$ can grow significantly with N, posing additional demands in hardware due to need of large adders. The remaining three algorithms do not suffer from these issues, as their sample-based comparison relies on simple operations yielding low scores.

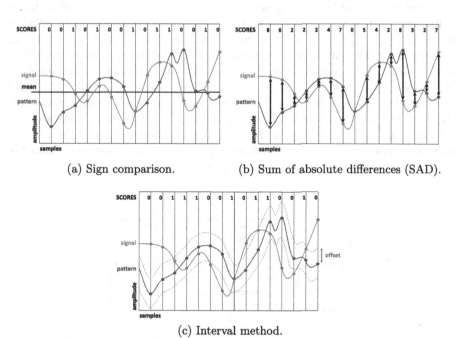

(a) Sign comparison.   (b) Sum of absolute differences (SAD).

(c) Interval method.

Fig. 1. Visualization of different algorithms for waveform matching.

We have ran several experiments in order to determine the suitability of all algorithms in the context of side channel signals, i.e. by testing their success rate in matching arbitrary patterns from real (noisy) measurements collected with an oscilloscope. Our experiments showed that the sign comparison algorithm has an unreliable triggering behaviour. Therefore we opt to discard it. SAD and interval matching algorithm perform rather well, and they both have the potential for low latency and good implementability. Flexibility is thus the criteria that determines our choice. In particular, the fact that the interval matching algorithm enjoys an extra degree of configurability via the offset. An additional benefit of interval matching is a better resistance to outliers. Note that for SAD, a large difference on a single sample may have a significant impact on the $T(k)$, potentially leading to a false negative. We therefore select interval matching as core algorithm for our design.

3 Architecture

In this section we describe the hardware architecture of our triggering system based on the interval matching algorithm. The top level view is shown in Fig. 2. The main components are a control unit, an analog-to-digital converter (ADC), and two modules responsible for the different modes of operation (capture and matching) sharing a memory block. The control unit provides a communication IO interface to enable external access for configuration. The incoming signal is first sampled by the ADC and then forwarded to the capture and matching modules. During capture mode, samples provided by the ADC are stored in memory when indicated by the capture signal. The amount of samples that can be stored is implementation-dependent, i.e. it is uniquely determined by the memory length. The captured measurement can be read through the IO interface. During matching mode, a pattern is first written to memory through the IO interface. The parameters of the algorithm (offset, threshold) are also externally set. The matching module contains an instantiation of the interval matching algorithm. If a match between the sampled incoming signal and the programmed reference is found, the trigger signal is activated.

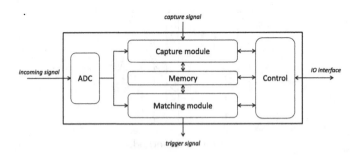

Fig. 2. Top level architecture view.

Due to its basic structure, the architecture of the capture module is not described in detail. Instead, we focus on the more critical matching module

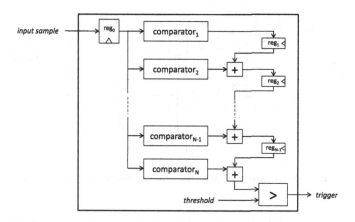

Fig. 3. Architecture of the matching module.

which determines the latency of the system. Its architecture, depicted in Fig. 3, is essentially a shifting integration structure formed by a parallel cluster of comparator blocks and a register adder chain that shifts its content towards the threshold comparison. Each comparator block checks whether an incoming sample lies in the interval specified by the reference and the offset for $g(m)$, i.e. it performs the test described in Eq. 4. The interval comparison is done in parallel for all samples in the pattern and thus we require N comparator blocks. The binary output of the comparator is fed to the register adder chain, which keeps track of the aggregated results from the previous comparisons. That way when a certain sample $h(m+k)$ is at reg_0, the result from comparing i previous samples is already stored at reg_i. This architecture enables a quick response time when a similar pattern appears in the side channel trace. The outcome of $comparator_N$ is added to reg_{N-1}, which contains the aggregated result of the previous $N-1$ samples. The resulting $T(k)$ is directly compared with the threshold to determine whether a trigger needs to be generated, i.e. in case of a match. This full step can be implemented in one clock cycle, and it is the critical path in the module. The latency of our architecture is 4 clock cycles, since it includes buffers in the IO pins and an extra register for the flexible hold-off time before triggering.

A more detailed view of the architecture of a comparator block and the following register adder chain is illustrated in Fig. 4. An input sample stored in reg_0 is compared against the upper and lower values of the interval for a sample $g(m)$. The pair $(upper\ limit_m, lower\ limit_m)$ is calculated in advance and stored in the corresponding registers UL_m and LL_m, i.e. UL_m contains $g(m)+$ offset and LL_m contains $g(m)-$ offset. The binary output of the interval comparisons is fed to an AND gate whose output determines whether the value in reg_{m-1} needs to be incremented before storing it in reg_m. The reset register allows to adjust the length of the reference signal, adding yet another flexibility feature to our design.

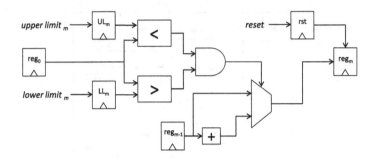

Fig. 4. Architecture of comparator block and register adder chain.

Note that the architecture given in Fig. 4 is slightly different for the first and last comparator blocks. In particular, the output of the AND gate in $comparator_1$ can be directly stored in reg_1. For the last comparator block, the output of the multiplexer goes to the threshold comparator.

4 Implementation

In this section we describe the realization of a trigger box based on the design and architecture of our waveform-matching based triggering system. We provide a brief description of the main components and interfaces and list the characteristics achieved after synthesis and place-and-route on an FPGA.

We have described our architecture in VHDL and implemented it on an Altera Cyclone IV GX FPGA Development Kit [2]. This low-end FPGA is equipped with 150k logic elements and 6.5 Mbits of embedded memory. Our hardware description deliberately avoids the use of any manufacturer-specific IP block, and thus can be easily ported to other commercial FPGAs. We have used Quartus II Web Edition Software for synthesis, place-and-route and programming. The Cyclone IV development board is not equipped with an ADC, required to sample the incoming side channel signal. Therefore we use an external Terasic AD/DA Data Conversion card [3]. This card provides two 14-bit ADC with a maximal sampling rate of 150 MS/s. It interfaces with the Cyclone IV development board via a standard HSMC interconnect header. A DC block is placed in front of the ADC to remove the DC component from the incoming signal.

We have enabled an RS-232 serial interface for IO communication and devised a rich instruction set to allow for external configuration. The commands allow at any time to select between capture or matching mode, read captured signals, program references, set the parameters of the interval matching (offset and threshold), vary the internal sample rate, and assign hold-off times for capturing signals and/or trigger generation, among others. Both capture and trigger signals are implemented as GPIO pins. For the latter we could alternatively use one of the digital-to-analog converters (DAC) of the Terasic AD/DA card.

Synthesis Results. The figures of our trigger box depend directly on the resources of the FPGA. We run the synthesis and place-and-route processes optimizing for speed. With this we obtain a design that can run at 171.17 MHz and allows for a pattern length of 1 500 samples. We clock the trigger box at 125 MHz using the built-in global oscillator, therefore obtaining a latency of 32 ns for our architecture. Taking into account the 96 ns delay caused by the ADC, the total latency of our trigger box is 128 ns. Note that by using Altera's specific IP blocks for PLLs we could generate a faster clock and thus slightly decrease our latency. The resource occupation is 133k logic elements (around 88 %) and 86k flip-flops. Most resources are occupied by the 14-bit comparators. The memory depth for trace capture is only limited by the memory resources of the FPGA, which is rather large. For the purposes of testing we fix it to 60 000 samples.

Note that several tradeoffs are possible. In particular, we can increase the maximum reference length by lowering the sample resolution. Dropping the least significant bits of the ADC output from 14 to 12 bits decreases the demand of the comparators. This allows to increase the reference length to 1 875 samples at the cost of some precision. Further reductions are also possible. The impact of tradeoffs on the performance of the trigger box can only be empirically evaluated, as it will naturally depend on the properties of the target device and the side channel signal.

For the sake of completeness, we list in Table 1 some features of our trigger box and the icWaves solution from Riscure. The latter figures are retrieved from the product data sheet [4]. Note that features related to signal conditioning are omitted from the listing in the table, as they will be discussed in Sect. 5. Other common features such as hold-off times are similarly unlisted.

Table 1. Main features of our trigger box designs and icWaves.

	Our design			icWaves
Algorithm	Interval matching			SAD
Latency	128 ns			500 ns
Sample rate	125 Msamples/s			200 Msamples/s
Resolution	14-bit	12-bit	8-bit	8-bit
Sample length	1 500	1 875	2 625	(1×512) or (2×256)
Memory depth	60k	60k	120k	8 000k

5 Evaluation

In this section we evaluate the performance of our trigger box by means of practical experiments. Our tests involve two different devices and two different types of side channel signals. We discuss the role of the interval matching parameters on the triggering behaviour and highlight the importance of analog signal conditioning.

5.1   Experimental Setup

Figure 5 depicts the main components of our experimental setup as well as their interconnections. The side channel signal of the cryptographic device is connected to an oscilloscope and to the trigger box. We use the version with 14-bit resolution in all experiments. Note that the signal is modified by means of analog circuitry before being fed to the ADC of the trigger box. This is required to map the signal amplitude to the input range of the ADC, i.e. to minimize quantization errors. Additional circuitry can be used to highlight interesting features of the signal, as will be discussed in the next section. We use a computer to operate the trigger box through the serial interface.

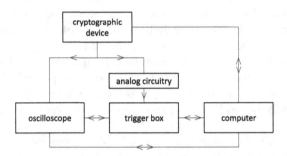

Fig. 5. Experimental setup.

Capture Mode. The computer begins by configuring the trigger box to enter capture mode. It then sends a command to the cryptographic device in order to start a cryptographic computation. The oscilloscope, which monitors the side channel signal of the cryptographic device, is responsible to indicate to the trigger box when to start recording. The incoming signal is then stored in internal memory (60 000 samples) and sent to the PC.

Note that in our setup the oscilloscope is responsible for activating the trigger box during capture mode. Hence it is required to (at least once) be able to configure the oscilloscope to trigger near the window of interest. This step can be done with the usual techniques, e.g. by detecting simple features on the IO line or on the side channel signal. The activation signal is provided by the AUX out port of the oscilloscope. An optional hold-off time can be specified either in the oscilloscope or in the trigger box. We stress that this step only needs to be executed once. During our experiments, we have always been able to carry it out without major difficulties, even if the trigger programmed in the scope has an unstable behaviour.

Matching Mode. The first step in matching mode consists in selecting a pattern from the pre-recorded trace. Any software visualization tool can be used for this, e.g. Matlab. The pattern is sent to the trigger box along with the parameters of the interval matching algorithm (offset and threshold). Once the trigger box is configured, the computer sends a command to the cryptographic device to start the cryptographic operations. If a match is found the trigger box generates a pulse on the trigger signal, optionally with a certain hold-off time. In our experimental setup this signal is connected to an analog channel of the oscilloscope. This allows us to easily verify whether the trigger box has found a matching pattern and, if so, whether it is the correct one. We note that in a real attack scenario the output trigger signal will be directly connected to an acquisition device to start measurement collection, or alternatively to some equipment for fault injection.

5.2 Experiments with a Java Card Smart Card

Our first evaluation of the trigger box is performed on a Java card that contains an applet to compute RSA signatures. The details of the implementation are unknown to us. We access the side channel power consumption of the Java card by measuring the voltage drop over a shunt resistor placed in the ground line. An exemplary trace collected with an oscilloscope at 80 MS/s is shown in Fig. 6 (top). The two long similar patterns indicate that the implementation uses RSA-CRT. Zooming into either of the CRT-branches reveals repetitive patterns corresponding to the inner modular operations, i.e. square and multiply. This is illustrated in Fig. 6 (bottom).

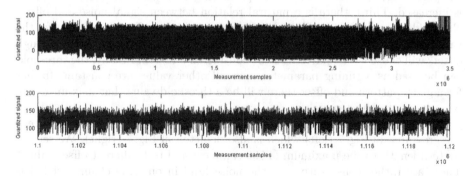

Fig. 6. Power measurements of the RSA signature applet in the Java card. Full execution (top), zoom into first CRT-branch (bottom).

Single Matching. We begin our experiments by bringing our setup to capture mode and configure it to record a trace close to the beginning of the first CRT-branch. The aim is to find a unique pattern that appears only *once* per cryptographic execution. We set the sampling rate of the trigger box at 125 MS/s. An exemplary recorded trace of 60 000 samples is shown in Fig. 7 (left). Note that the shape of the trace in Fig. 7 differs from the one in Fig. 6. This is due to the ADC in the trigger box being different from the ADC in the oscilloscope.

The next steps are to select a pattern and to set the parameters of the interval matching algorithm. The behaviour of the trigger box will naturally depend on

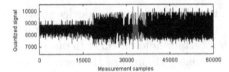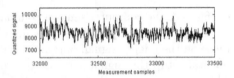

Fig. 7. Trace recorded by the trigger box during capture mode, pattern selected for matching mode in grey (left), zoom into the pattern (right).

a good combination of these variables. Our goal is to obtain a single correct trigger per execution while avoiding (or at least minimizing) false negative and particularly false positives.

It is important to stress that there is no generic rule on how parameters need to be set. However, some general observations can be made. First and most important, it is critical that the pattern we want to trigger on is as unique as possible. Since the length of the reference is limited, a way to incorporate more unique features is by varying the sample rate. This variation should however not result in an undersampling of the signal. Second, using the maximal pattern length is in general beneficial for detection. However, once the unique features are incorporated in the pattern it may not make a difference in terms of success rate. We have empirically observed the existence of a convergence point in which the percentage of correct matches stops varying even when the reference length is increased. Third, there is a natural relation between signal noise and offset. Noisy side channel signals will inevitably demand higher offsets to account for the variability in the observations. This comes however at a risk, as arbitrarily high offsets will cause wrong patterns to be considered a match. Finally, the threshold can be used as a tuning parameter once all other values are constant. In fact for a given pattern and offset there will be a threshold value that maximizes the percentage of correct matches.

Let us illustrate this behaviour with an example. Assume we select as pattern the grey area in Fig. 7 (left) which is also depicted in Fig. 7 (right). We set the pattern length to the maximum (1 500 samples) and test different offset values. These are rather conservative, as the noise level in our side channel signal is quite low. Figure 8 shows an area plot that visualizes the responses of the trigger box for each offset value in function of the threshold. The vertical axes shows the outcome (in percentage) computed by running multiple experiments. Each color (grey, black, white) indicates a possible outcome of the experiments (false positive, correct match, false negative).

The plots clearly show that for a given offset the desired triggering behaviour can be tuned by varying the threshold. In most cases one wants to avoid false positives. Therefore, it is preferable to select a threshold that is slightly higher than the one that maximizes the percentage of correct matches, i.e. to move away from the grey area in the plot. We also observe that higher offsets demand higher thresholds in order to avoid false positives. In contrast, the maximal percentage of correct matches only varies slightly with the offset.

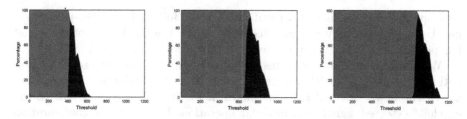

Fig. 8. Percentage of outcomes for varying thresholds: grey (false positive), black (correct match), white (false negative). Values of offset are 150 (left), 300 (middle), 450 (right).

Our best results when triggering at the beginning of the first CRT-branch are obtained when selecting the combination offset of 450 and threshold of 900, which yields a percentage of correct matches of 90 %, with 10 % false negatives and no false positives.

Multiple Matching. Our follow-up experiment consists in checking whether it is possible to trigger on *each* execution of a modular operation in the CRT-branches. For this we need to select the repetitive pattern in Fig. 6 (bottom), which perfectly characterizes the occurrence of squarings and multiplications. We lower the sampling rate of the trigger box to 62.5 MS/s in order to fit the pattern in less than the maximum 1 500 samples supported by our implementation. The recorded pattern is shown in Fig. 9 (left).

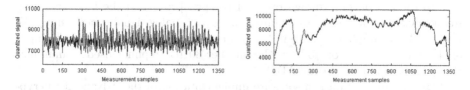

Fig. 9. Repetitive pattern in CRT-branches. Normal (left), with envelope detection (right).

Although the pattern appears to contain characteristic features, it turns out to be completely unsuitable for the purposes of triggering. In fact, all our experiments yield a negligible percentage of correct matches for any combination of offset and threshold. Most observed outcomes are false negatives or false positives. The reason behind these results is the repetitiveness of the features in the reference, which is caused by the relatively high frequency of the side channel signal. This causes a non unique pattern with many matches in the incoming signal.

In order to overcome this issue, some extra analog circuitry is required to condition the signal before the ADC and highlight its low-frequency features. We opt to incorporate an ADL5511 envelope detector board with some added capacitors that effectively reduces the bandwidth of the incoming signal to 2 MHz. The envelope detector consists of a rectification stage followed by a low-pass filtering. Using this circuitry we can effectively reduce the bandwidth of the signal independently of its high frequencies, and therefore capture useful features.

We have repeated our experiments by using an envelope detector in combination with an extra 30 dB amplifier (Langer PA 303) to minimize quantization noise. The newly selected pattern, shown in Fig. 9 (right), corresponds to the envelope detected signal of a modular operation. By using this configuration with an offset of 50 and a threshold of 810, we are able to increase our success rate to nearly 100 % without any false positive, i.e. our box is able to trigger on almost every occurrence of a modular operation in a single RSA-CRT operation. In fact, the only false negatives correspond to the first operation in each CRT-branch, which have a shape different than the rest.

5.3 Experiments with an ARM Processor

For our second evaluation of the trigger box we switch to a more challenging platform featuring non-deterministic program execution. We use a BeagleBone Black [1] equipped with a Sitara ARM Cortex-A8 32-bit RISC processor. The non-deterministic behaviour of the processor stems from its dynamic branch predictors, L1 and L2 cache memories and out-of-order execution engine. Additionally, the processor runs a preemptive multitasking Linux operating system with e.g. scheduling, context switches and interrupts. The core runs at a clock frequency of 1 GHz.

In order to monitor the power consumption, we place an electromagnetic pen probe on one of the decoupling capacitors as previously described in [6]. Due to the low output voltage range of the pen probe and the high operating frequency of the processor, we require some extra analog signal conditioning. We insert a 30 dB amplifier (Langer PA 303) between the pen probe and the input of the envelope detector, followed by another 30 dB amplifier used to minimize the quantization noise in the ADC.

Our test application is a software implementation of the Advanced Encryption Standard (AES) [12] that performs bulk encryption. Finding a consistent and reliable trigger on the side channel signal is an arduous task when using only an oscilloscope, i.e. sudden amplitude changes occur frequently due to other running processes causing multiple false positives. However, we only need one correct trigger in order to store a measurement during capture mode.

A trace recorded by our trigger box is shown in Fig. 10 (left). Thanks to the envelope detector, the batched executions of the AES are identifiable in the trace (30 encryptions in this particular case). We select a pattern corresponding to a single execution, depicted in Fig. 10 (right), and program it into the trigger box. The length of this reference signal is 960 samples, and it is obtained with a sampling rate of 25 MS/s. The lower sampling rate is an additional benefit from using an envelope detector, as it allows to sample the now low-frequency signal without loosing its features.

By setting the offset to 130 and choosing a threshold of 630, we achieve 99.2 % of correct matches, i.e. we trigger on almost all AES executions, without any false

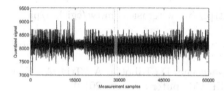 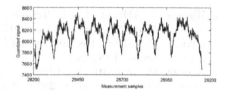

Fig. 10. Trace recorded by the trigger box during capture mode, pattern selected for matching mode in grey (left), zoom into the pattern (right).

positive. The remaining 0.8 % of false negatives correspond to the first encryption and any other execution occurring after an interrupt. This is an artifact of the instruction cache of the processor, which is filled with code after every context switch. The cache filling leads to a different shape in the side channel signal, which the trigger box correctly identifies as no match.

It is worth mentioning that our success rate is kept even when we increase the number of batched encryptions to e.g. 50 000. The same outcome is achieved when we increase the CPU load to 100 % by enabling heavy computational tasks in parallel such as RSA key generation in OpenSSL.

6   Conclusions

Triggering is critical to enable implementation attacks in real scenarios. The lack of accurate and reliable trigger points may fully prevent to mount an attack on a certain implementation, even if unprotected. This is particularly true for modern high-end devices with non-deterministic behaviour due to complex CPU architecture and operating system. In this work we have documented the design, architecture, implementation and evaluation of a waveform-matching based triggering system tailored to the context of embedded security. With this, we put forward a tool that can be of use to the research community.

Acknowledgements. We would like to thank Victor Förster for initial contributions to the system's design and architecture. This work was supported in part by the Research Council KU Leuven: C16/15/058. In addition, this work was supported by the Flemish Government, FWO G.0550.12N, by the Hercules Foundation AKUL/11/19, and through the Horizon 2020 research and innovation programme under grant agreement 644052 HECTOR. Benedikt Gierlichs is a Postdoctoral Fellow of the Fund for Scientific Research - Flanders (FWO).

References

1. BeagleBone Black Starting Guide. Beagleboard.org. http://beagleboard.org/ getting-started. Accessed Dec 2015
2. Cyclone IV GX FPGA Development Kit. Altera. https://www.altera.com/ products/boards_and_kits/dev-kits/altera/kit-cyclone-iv-gx.html. Accessed Dec 2015

3. Highspeed AD/DA Card. Terasic. http://www.terasic.com.tw/cgi-bin/page/archive.pl?No=278. Accessed Dec 2015
4. icWaves Datasheet. Riscure. https://www.riscure.com/security-tools/hardware/icwaves. Accessed Dec 2015
5. Agoyan, M., Dutertre, J., Mirbaha, A., Naccache, D., Ribotta, A., Tria, A.: How to flip a bit? In: IOLTS, pp. 235–239. IEEE Computer Society (2010)
6. Balasch, J., Gierlichs, B., Reparaz, O., Verbauwhede, I.: DPA, bitslicing and masking at 1 GHz. In: Güneysu, T., Handschuh, H. (eds.) CHES 2015. LNCS, vol. 9293, pp. 599–619. Springer, Heidelberg (2015)
7. Balasch, J., Gierlichs, B., Verbauwhede, I.: An In-depth and black-box characterization of the effects of clock glitches on 8-bit MCUs. In: Breveglieri, L., Guilley, S., Koren, I., Naccache, D., Takahashi, J. (eds.) FDTC, pp. 105–114. IEEE Computer Society (2011)
8. Bar-El, H., Choukri, H., Naccache, D., Tunstall, M., Whelan, C.: The sorcerer's apprentice guide to fault attacks. Proc. IEEE 94(2), 370–382 (2006)
9. Boneh, D., DeMillo, R.A., Lipton, R.J.: On the importance of checking cryptographic protocols for faults. In: Fumy, W. (ed.) EUROCRYPT 1997. LNCS, vol. 1233, pp. 37–51. Springer, Heidelberg (1997)
10. Clavier, C., Coron, J.-S., Dabbous, N.: Differential power analysis in the presence of hardware countermeasures. In: Paar, C., Koç, Ç.K. (eds.) CHES 2000. LNCS, vol. 1965, p. 252. Springer, Heidelberg (2000)
11. Coron, J.-S., Kizhvatov, I.: An efficient method for random delay generation in embedded software. In: Clavier, C., Gaj, K. (eds.) CHES 2009. LNCS, vol. 5747, pp. 156–170. Springer, Heidelberg (2009)
12. Daemen, J., Rijmen, V.: The Design of Rijndael: AES - The Advanced Encryption Standard. Information Security and Cryptography. Springer, Heidelberg (2002)
13. Dehbaoui, A., Dutertre, J., Robisson, B., Tria, A.: Electromagnetic transient faults injection on a hardware and a software implementations of AES. In: Bertoni, G., Gierlichs, B. (eds.) FDTC, pp. 7–15. IEEE Computer Society (2012)
14. Gandolfi, K., Mourtel, C., Olivier, F.: Electromagnetic analysis: concrete results. In: Koç, Ç.K., Naccache, D., Paar, C. (eds.) CHES 2001. LNCS, vol. 2162, p. 251. Springer, Heidelberg (2001)
15. Kocher, P.C.: Timing attacks on implementations of Diffie-Hellman, RSA, DSS, and other systems. In: Koblitz, N. (ed.) CRYPTO 1996. LNCS, vol. 1109, pp. 104–113. Springer, Heidelberg (1996)
16. Kocher, P.C., Jaffe, J., Jun, B.: Differential power analysis. In: Wiener, M. (ed.) CRYPTO 1999. LNCS, vol. 1666, p. 388. Springer, Heidelberg (1999)
17. Quisquater, J.-J., Samyde, D.: ElectroMagnetic Analysis (EMA): measures and counter-measures for smart cards. In: Attali, S., Jensen, T. (eds.) E-smart 2001. LNCS, vol. 2140, p. 200. Springer, Heidelberg (2001)
18. Quisquater, J.-J., Samyde, D.: Eddy current for magnetic analysis with active sensor. In: Esmart 2002, pp. 185–194 (2002)
19. Schmidt, J., Herbst, C.: A practical fault attack on square and multiply. In: Breveglieri, L., Gueron, S., Koren, I., Naccache, D., Seifert, J. (eds.) FDTC, pp. 53–58. IEEE Computer Society (2008)
20. Skorobogatov, S.P., Anderson, R.J.: Optical fault induction attacks. In: Kaliski Jr., B.S., Koç, Ç.K., Paar, C. (eds.) CHES 2002. LNCS, vol. 2523, pp. 2–12. Springer, Heidelberg (2003)
21. van Woudenberg, J.G.J., Witteman, M.F., Menarini, F.: Practical optical fault injection on secure microcontrollers. In: Breveglieri, L., Guilley, S., Koren, I., Naccache, D., Takahashi, J. (eds.) FDTC, pp. 91–99. IEEE Computer Society (2011)

Robust and One-Pass Parallel Computation of Correlation-Based Attacks at Arbitrary Order

Tobias Schneider[1(✉)], Amir Moradi[1], and Tim Güneysu[2]

[1] Horst Görtz Institute for IT Security,
Ruhr-Universität Bochum, Bochum, Germany
{tobias.schneider-a7a,amir.moradi}@rub.de
[2] University of Bremen and DFKI, Bremen, Germany
tim.gueneysu@uni-bremen.de

Abstract. The protection of cryptographic implementations against higher-order attacks has risen to an important topic in the side-channel community after the advent of enhanced measurement equipment that enables the capture of millions of power traces in reasonably short time. However, the preprocessing of multi-million traces for such an attack is still challenging, in particular when in the case of (multivariate) higher-order attacks all traces need to be parsed at least two times. Even worse, partitioning the captured traces into smaller groups to parallelize computations is hardly possible with current techniques.

In this work we introduce procedures that allow iterative computation of correlation in a side-channel analysis attack at any arbitrary order in both univariate and multivariate settings. The advantages of our proposed solutions are manifold: (i) they provide stable results, i.e., by increasing the number of used traces high accuracy of the estimations is still maintained, (ii) each trace needs to be processed only once and at any time the result of the attack can be obtained (without requiring to reparse the whole trace pool when adding more traces), (iii) the computations can be efficiently parallelized, e.g., by splitting the trace pool into smaller subsets and processing each by a single thread on a multi-threading or cloud-computing platform, and (iv) the computations can be run in parallel to the measurement phase. In short, our constructions allow efficiently performing higher-order side-channel analysis attacks (e.g., on hundreds of million traces) which is of crucial importance when practical evaluation of the masking schemes need to be performed.

1 Introduction

Side-channel analysis (SCA) poses a major threat for security-sensitive applications. This becomes particularly critical when the cryptographic device – particularly in pervasive applications – is delivered to the end user, where it is operated in a hostile environment (cf. [17,22]). For such a case the integration of appropriate countermeasures against SCA attacks has become essential in the design of the device. In this context, *masking* as a countermeasure obtained the most attraction from both academia and industry due to its sound theoretical

© Springer International Publishing Switzerland 2016
F.-X. Standaert and E. Oswald (Eds.): COSADE 2016, LNCS 9689, pp. 199–217, 2016.
DOI: 10.1007/978-3-319-43283-0_12

basis as well as its practical efficiency to mitigate the attacks. Masking counter-measures are based on the principle of *secret sharing* for which many different forms including Boolean, arithmetic, multiplicative, polynomial base, etc. have been proposed [5,6,19].

Since the efficiency of a masking schemes strongly depends on its implementation, a practical evaluation of the final product (or a prototype) is inevitable. For this situation, techniques such as the *test vector leakage assessment* [9] (known as t-test) have been developed to practically examine the vulnerability of a cryptographic design. However, such an evaluation scheme can only report the *existence* of a leakage in a product, but it does not provide any indication whether this leakage is indeed *exploitable* by an attack. In reply to the question if a leakage is in fact exploitable for key recovery, one needs to mount different SCA attacks and examine their success. Depending on the definition and settings of the masking scheme, it can provide security against SCA attacks up to a certain order d. Consequently, all tests and attacks need to take all particular orders ranging from 1 up to $d + 1$ into account.

The most common SCA attack, Correlation Power Analysis (CPA) [4], is based on a hypothetical leakage model and the estimation of correlation (commonly by Pearson's correlation coefficient) between the hypothetical leakages and the SCA traces. In its simplest setting, the attack runs independently at each sample point of the SCA traces. This univariate first-order CPA can be extended to higher orders $d > 1$ by introducing a preprocessing stage for the traces at each sample point. This preprocessing involves the computation of mean-free values which are then squared (for a univariate $d = $ 2nd-order CPA), cubed (for a univariate $d = $ 3rd-order CPA), or any corresponding power for larger d. Prior to the attack d different sample points of each trace are combined into a centered product for the multivariate case at order $d > 1$. In other words, first mean-free representations are calculated of which d sample points of each trace are multiplied. It is noteworthy that finding such d points of interest is another challenging task which has been well studied in [8,18].

By increasing the order of the underlying masking scheme the corresponding higher-order CPA becomes more susceptible to noise. Indeed the number of required traces to mount a successful attack increases exponentially in d with respect to the noise standard deviation. Therefore, a higher-order attack typically requires several (hundreds of) millions of traces to be successful [1,13]. The conventional strategy for preprocessing (known as "three-pass") parses all traces three times to (i) obtain the means, (ii) combine the desired points by their mean-free product, and (iii) estimate the correlation[1]. This procedure has many shortcomings as by adding more traces to the trace pool, the entire last two steps need to be repeated. Hence, it is not easily possible to parallelize the computations by splitting the trace pool into smaller sets. We should emphasize that, in case of univariate attacks, the parallelization can be trivially done by splitting each trace into smaller subtraces with a lower number of sample points.

---

[1] In some particular cases, e.g., univariate, the last two steps can be combined.

Alternatively as shown in [3] for first-order and second-order CPA, the formulas for preprocessing and the estimation of the correlation can be combined by following the displacement law. This procedure (so-called "Raw-Moment") solves all the shortcomings of the three-pass approach. In fact:

- When increasing the trace pool, the estimated raw moments are easily updated by only processing the given new traces.
- The attack can be started before the measurement phase is completed. This helps to further increase the performance of the attacks.
- The result of the attack can be obtained without introducing any overhead to the process of the further traces at any time during the measurement phase.
- The trace pool can be easily split into smaller sets and each set can be processed independently by different threads. Due to the nature of the raw moments, the result of different threads (at any time) can be easily combined to derive the result of the attack.

Note, however, that this procedure was only presented for first-order and bivariate second-order CPA using 10,000,000 traces and may suffer from numerical instabilities as the raw moments become pretty large values by increasing the number of traces. Hence, it can lead to serious accuracy loss due to the limited fraction significand of floating point formats (e.g., IEEE 754). This issue becomes extremely problematic for higher-order $(d > 2)$ attacks.

The instability in formulas that are based on raw moments has been previously studied to come up for appropriate solutions. For example, in [15] robust iterative formulas for centralized and standardized moments at any arbitrary order as well as for correlation are given that avoid such instabilities by increasing the number of samples. Furthermore, iterative formulas for the t-test at any arbitrary order are given in [20].

Our Contribution: In this work, we present an approach based on centralized and standardized moments to cover univariate as well as multivariate CPA attacks at any arbitrary order. Our solution benefits from all the aforementioned advantages of the raw-moment approach while it maintains the accuracy (as for the three-pass approach) regardless of the order of the attack and the number of traces. This work not only covers CPA attacks but also Moments-Correlating DPA [14] where moments are correlated to the (preprocessed) traces with the goal of avoiding the necessity of a hypothetical leakage model (that is unavoidable in CPA attacks).

Prior to the description of our solution we define two terms *iterative* and *incremental* which are frequently used in the rest of the paper. Suppose that after finishing all the required processes on the trace pool $\mathcal{Q}$, a new trace y is added to the trace pool $\mathcal{Q}' = \mathcal{Q} \cup \{y\}$. We provide *incremental* formulas that allow updating the previously computed terms by only processing the new trace y. In addition to that, we suppose that the trace pool $\mathcal{Q}$ is divided into two groups as $\mathcal{Q} = \mathcal{Q}_1 \cup \mathcal{Q}_2$, and each group is independently processed using the given incremental formulas. We provide (two-pair) *iterative* formulas that enable the combination of results computed over each group $\mathcal{Q}_1$ and $\mathcal{Q}_2$ to derive the result of the full trace pool $\mathcal{Q}$.

2   Notations

We use capital letters for random variables, and lower-case letters for their realizations. Vectors are denoted with bold notations, functions with sans serif fonts, and sets with calligraphic ones.

Suppose that in a side-channel attack, with respect to n queries with associated data (e.g., plaintext or ciphertext) $\boldsymbol{d}_{i \in \{1,\ldots,n\}}$, n side-channel measurements (so-called traces) are collected. Let us denote each trace by $\boldsymbol{t}_{i \in \{1,\ldots,n\}}$ containing m sample points $\{t_i^{(1)}, \ldots, t_i^{(m)}\}$.

Following the divide-and-conquer principle, one objective of a side-channel attack is to recover a part k of the secret key $\boldsymbol{k}$, which contributed to the processing of the entire associated data $\boldsymbol{d}_{i \in \{1,\ldots,n\}}$. Prior to the attack an intermediate value V is selected, which given the associated data and a key guess k is predictable, i.e., $v_i = \mathsf{F}(\boldsymbol{d}_i, k)$. In a CPA attack a hypothetical leakage model $\widetilde{\mathsf{L}}(.)$ is applied on the chosen intermediate value which should be (sufficiently) linearly proportional to the actual leakage of the target device, i.e., $\mathsf{L}(.)$. As a common and straightforward example, the Hamming weight of an Sbox output during the first round of an encryption function is employed when attacking an exemplary micro-processor based implementation, i.e., $l_i = \widetilde{\mathsf{L}}(v_i) = HW(\mathsf{S}(d_i \oplus k))$, where d_i denotes a necessary part of $\boldsymbol{d}_i$ to predict v_i.

Let us denote the dth-order raw statistical moment of a random variable X by $M_d = \mathsf{E}(X^d)$, with $\mu = M_1$ the mean and $\mathsf{E}(.)$ the expectation operator. We also denote the dth-order ($d > 1$) central moment by $CM_d = \mathsf{E}\left((X - \mu)^d\right)$, with $s^2 = CM_2$ the variance. Finally, the dth-order ($d > 2$) standardized moment is denoted by $SM_d = \mathsf{E}\left(\left(\frac{X-\mu}{s}\right)^d\right)$, with SM_3 the skewness and SM_4 the kurtosis.

3   Univariate CPA

For a *univariate* CPA attack the correlation between the traces $\boldsymbol{T}$ and the hypothetical leakage values L is estimated. Due to the *univariate* nature of the attack, such a process is performed at each sample point $(1, \ldots, m)$ independently. Therefore, below – for simplicity – we omit the upper index of the sample points and denote a sample point of the ith trace by t_i.

The estimation of the correlation with Pearson correlation coefficient (as the normalized covariance) is defined as

$$\rho = \frac{\mathrm{cov}(T, L)}{s_t \, s_l} = \frac{\mathsf{E}\left((T - \mu_t)(L - \mu_l)\right)}{s_t \, s_l}, \tag{1}$$

where μ_t (resp. μ_l) denotes the estimated mean of the traces (resp. of the hypothetical leakages). s_t (resp. s_l) also stands for standard deviation.

In the discrete domain we can write

$$\rho = \frac{\dfrac{1}{n}\sum_{i=1}^{n}(t_i - \mu_t)(l_i - \mu_l)}{\sqrt{\dfrac{1}{n}\sum_{i=1}^{n}(t_i - \mu_t)^2 \dfrac{1}{n}\sum_{i=1}^{n}(l_i - \mu_l)^2}} \tag{2}$$

Based on the way followed in [3] one can write

$$\rho = \frac{\dfrac{1}{n}\sum_{i=1}^{n} t_i\, l_i - \mu_t\, \mu_l}{\sqrt{\left(\dfrac{1}{n}\sum_{i=1}^{n} t_i^2 - \mu_t^2\right)\left(\dfrac{1}{n}\sum_{i=1}^{n} l_i^2 - \mu_l^2\right)}} = \frac{M_{1,\mathcal{T}\cdot\mathcal{L}} - M_{1,\mathcal{T}}\, M_{1,\mathcal{L}}}{\sqrt{\left(M_{2,\mathcal{T}} - M_{1,\mathcal{T}}^2\right)\left(M_{2,\mathcal{L}} - M_{1,\mathcal{L}}^2\right)}}, \tag{3}$$

which are based on dth-order raw moments, i.e., $M_{d,\mathcal{X}} = \frac{1}{n}\sum_{i=1}^{n} x_i^d$. However, as stated in [20], such constructions can lead to numerically unstable situations [10]. During the computation of the raw moments the intermediate values tend to become very large which can lead to a loss in accuracy. Further, M_2 and M_1^2 can be large values, and the result of $M_2 - M_1^2$ can also lead to a significant accuracy loss due to the limited fraction significand of floating point formats (e.g., IEEE 754).

Iterative. We can alternatively write

$$\rho = \frac{\dfrac{1}{n}\sum_{i=1}^{n}(t_i - \mu_t)(l_i - \mu_l)}{\sqrt{\dfrac{1}{n}\sum_{i=1}^{n}(t_i - \mu_t)^2 \dfrac{1}{n}\sum_{i=1}^{n}(l_i - \mu_l)^2}} = \frac{\dfrac{1}{n} ACS_1}{\sqrt{\dfrac{1}{n} CS_{2,\mathcal{T}} \dfrac{1}{n} CS_{2,\mathcal{L}}}}, \tag{4}$$

with $CS_{d,\mathcal{X}} = \sum_{i=1}^{n}(x_i - \mu_x)^d$ as the definition of dth-order *centralized sum* given in [20]. Further, we define ACS_1 as the first-order *adjusted centralized sum*.

Suppose that $M_{1,\mathcal{Q}_1}$ (resp. $M_{1,\mathcal{Q}_2}$) denotes the first raw moment (sample mean) of the given set $\mathcal{Q}_1$ (resp. $\mathcal{Q}_2$) with cardinality $n_1 = |\mathcal{Q}_1|$ and $n_2 = |\mathcal{Q}_2|$. $M_{1,\mathcal{Q}}$ as the first raw moment of $\mathcal{Q} = \mathcal{Q}_1 \cup \mathcal{Q}_2$ can be written as [15]

$$M_{1,\mathcal{Q}} = \frac{n_1\, M_{1,\mathcal{Q}_1} + n_2\, M_{1,\mathcal{Q}_2}}{n}, \tag{5}$$

with $n = n_1 + n_2$ as the cardinality of $\mathcal{Q}$.

In the same way, such a formula can be written for the centralized sum $CS_{d,\mathcal{Q}}$ at any arbitrary order $d > 1$ as [15]

$$CS_{d,\mathcal{Q}} = CS_{d,\mathcal{Q}_1} + CS_{d,\mathcal{Q}_2} + \sum_{p=1}^{d-2} \binom{d}{p} \left[\left(\frac{-n_2}{n}\right)^p CS_{d-p,\mathcal{Q}_1} + \left(\frac{n_1}{n}\right)^p CS_{d-p,\mathcal{Q}_2} \right] \Delta^p$$
$$+ \left(\frac{n_1 n_2}{n} \Delta\right)^d \left[\left(\frac{1}{n_2}\right)^{d-1} - \left(\frac{-1}{n_1}\right)^{d-1} \right],$$
$$(6)$$

with $\Delta = M_{1,\mathcal{Q}_2} - M_{1,\mathcal{Q}_1}$. It is noteworthy that the calculation of $CS_{d,\mathcal{Q}}$ additionally requires $CS_{p,\mathcal{Q}_1}$ and $CS_{p,\mathcal{Q}_2}$ for $1 < p \le d$.

The remaining part is the first-order adjusted centralized sum ACS_1. Suppose that $\mathcal{Q}_1$ and $\mathcal{Q}_2$ denote sets of doubles (t, l) with first-order adjusted centralized sum $ACS_{1,\mathcal{Q}_1}$ and $ACS_{1,\mathcal{Q}_2}$ respectively. The first-order adjusted centralized sum of $\mathcal{Q} = \mathcal{Q}_1 \cup \mathcal{Q}_2$ can be written as

$$ACS_{1,\mathcal{Q}} = ACS_{1,\mathcal{Q}_1} + ACS_{1,\mathcal{Q}_2} + \frac{n_1 n_2}{n} \Delta_t \Delta_l, \qquad (7)$$

with $\Delta_t = \mu_{t,\mathcal{Q}_2} - \mu_{t,\mathcal{Q}_1}$ and $\Delta_l = \mu_{l,\mathcal{Q}_2} - \mu_{l,\mathcal{Q}_1}$. For simplicity, we denote $M_{1,\mathcal{T}_1}$ by $\mu_{t,\mathcal{Q}_1}$ and $M_{1,\mathcal{L}_1}$ by $\mu_{l,\mathcal{Q}_1}$. The sets $\mathcal{T}_1$ and $\mathcal{L}_1$ are formed respectively from the first and second elements of the doubles in $\mathcal{Q}_1$ (the same holds for $\mathcal{Q}_2$, $\mu_{t,\mathcal{Q}_2}$, and $\mu_{l,\mathcal{Q}_2}$).

Incremental, $n_2 = 1$. We now optimize the computations of each set. It is indeed enough to suppose that $\mathcal{Q}_2$ consists of only one element y. Hence the update formula for the first raw moment can be written as

$$M_{1,\mathcal{Q}} = M_{1,\mathcal{Q}_1} + \frac{\Delta}{n},$$

with $\Delta = y - M_{1,\mathcal{Q}_1}$. Note that $\mathcal{Q}_1$ and $M_{1,\mathcal{Q}_1}$ are initialized with $\emptyset$ and respectively zero. Similarly, we can write the same for the dth-order centralized sum

$$CS_{d,\mathcal{Q}} = CS_{d,\mathcal{Q}_1} + \sum_{p=1}^{d-2} \binom{d}{p} CS_{d-p,\mathcal{Q}_1} \left(\frac{-\Delta}{n}\right)^p + \left(\frac{n-1}{n}\Delta\right)^d \left[1 - \left(\frac{-1}{n-1}\right)^{d-1}\right],$$
$$(8)$$

where $\Delta = y - M_{1,\mathcal{Q}_1}$. For the first-order adjusted centralized sum we can also write

$$ACS_{1,\mathcal{Q}} = ACS_{1,\mathcal{Q}_1} + \frac{n-1}{n} \Delta_t \Delta_l, \qquad (9)$$

with $\Delta_t = t_n - \mu_{t,\mathcal{Q}_1}$ and $\Delta_l = l_n - \mu_{l,\mathcal{Q}_1}$, where $\mathcal{Q}_2 = \{(t_n, l_n)\}$.

Based on these formulas the correlation can be computed efficiently in one pass. Furthermore, since the intermediate results of the central sums are mean-free, they do not become significantly large which helps preventing the numerical instabilities.

3.1 Univariate Higher-Order CPA

Higher-order attacks require that the sample traces are preprocessed. For the second-order univariate CPA the preprocessing consists of making each sample point mean-free squared:

$$t_i' = (t_i - \mu_t)^2 \,.$$

For higher orders $d > 2$ the traces are usually additionally standardized as $\dfrac{t_i'}{s_t{}^d}$, where s_t denotes the standard deviation. Therefore, the Pearson correlation can be written as

$$\rho = \frac{\frac{1}{n} \sum\limits_{i=1}^{n} \left(\frac{t_i'}{s_t{}^d} - \frac{\mu_{t'}}{s_t{}^d} \right)(l_i - \mu_l)}{\sqrt{\frac{1}{n} \sum\limits_{i=1}^{n} \left(\frac{t_i'}{s_t{}^d} - \frac{\mu_{t'}}{s_t{}^d} \right)^2 \frac{1}{n} \sum\limits_{i=1}^{n} (l_i - \mu_l)^2}} = \frac{\frac{1}{n} \sum\limits_{i=1}^{n} \left(t_i'(l_i - \mu_l) \right)}{\sqrt{\frac{1}{n} \sum\limits_{i=1}^{n} (t_i' - \mu_{t'})^2 \frac{1}{n} \sum\limits_{i=1}^{n} (l_i - \mu_l)^2}} \tag{10}$$

The straightforward way is to first preprocess the entire trace set $t_{i \in \{1,\dots,n\}}$. Hence the measurement phase has to be completed before the preprocessing can be started. Another drawback is the reduced efficiency as each of the preprocessing and the estimation of the correlation steps needs at least one pass over the whole trace set.

In [3], the authors propose iterative formulas for first- and second-order CPA. Their approach is based on raw moments which can lead to numerical instability if the values get too large [20]. Alternatively, we propose an iterative method which is based on the centralized moments. These values are mean-free which leads to smaller values and better accuracy for a large number of measurements. This approach can be run in parallel to the measurements (and can be also split into smaller threads) as the result is incrementally updated for each new measurement. Therefore, it needs only one pass over the whole trace set. In the following, we present all necessary iterative formulas to perform a univariate CPA at any arbitrary order with sufficient accuracy. We divide the expressions by the numerator and denominator of Eq. (10).

3.2 Numerator

Note that even though the numerator looks similar to a raw-moment approach, it operates with centralized (mean-free) values. Therefore, numerical instabilities are avoided. The numerator for the d-th order correlation can be written as

$$\frac{1}{n} \sum_{i=1}^{n} \left(t_i'(l_i - \mu_l) \right) = \frac{1}{n} \sum_{i=1}^{n} (t_i - \mu_t)^d \, (l_i - \mu_l) = \frac{1}{n} ACS_d, \tag{11}$$

with ACS_d which we refer to as the dth-order *adjusted centralized sum*.

We start with a generic formula which merges the adjusted centralized sum of two sets $\mathcal{Q} \cup \mathcal{Q}_2 = \mathcal{Q}$ with $|\mathcal{Q}_1| = n_1$, $|\mathcal{Q}_2| = n_2$ and $|\mathcal{Q}| = n$. The goal is to compute $ACS_{d,\mathcal{Q}}$ given only the adjusted and centralized sums of $\mathcal{Q}_1$ and $\mathcal{Q}_2$.

Theorem 1. *Let $\mathcal{Q}_1$ and $\mathcal{Q}_2$ be given sets of doubles (t, l). Suppose also $\mathcal{T}_1$ and $\mathcal{L}_1$ as the sets of respectively the first and second elements of the doubles in $\mathcal{Q}_1$ (the same for $\mathcal{T}_2$ and $\mathcal{L}_2$). The dth-order adjusted centralized sum $ACS_{d,\mathcal{Q}}$ of the extended set $\mathcal{Q} = \mathcal{Q}_1 \cup \mathcal{Q}_2$ with $\Delta_t = \mu_{t,\mathcal{Q}_2} - \mu_{t,\mathcal{Q}_1}$ and $\Delta_l = \mu_{l,\mathcal{Q}_2} - \mu_{l,\mathcal{Q}_1}$ can be written as*

$$ACS_{d,\mathcal{Q}} = ACS_{d,\mathcal{Q}_1} + ACS_{d,\mathcal{Q}_2} + \frac{\Delta_l}{n}\left(n_1\, CS_{d,\mathcal{Q}_2} - n_2\, CS_{d,\mathcal{Q}_1}\right)$$

$$+ \sum_{p=1}^{d-1}\binom{d}{p}\left(\frac{\Delta_t}{n}\right)^p\left[(-n_2)^p\, ACS_{d-p,\mathcal{Q}_1} + (n_1)^p\, ACS_{d-p,\mathcal{Q}_2}\right.$$

$$+ \frac{\Delta_l}{n}\left.\left((-n_2)^{p+1}\, CS_{d-p,\mathcal{Q}_1} + (n_1)^{p+1}\, CS_{d-p,\mathcal{Q}_2}\right)\right]$$

$$+ \frac{\left(n_1\,(-n_2)^{d+1} + n_2\,(n_1)^{d+1}\right)}{n^{d+1}}(\Delta_t)^d\,\Delta_l \tag{12}$$

The proof of Theorem 1 is omitted due to length restrictions.

Incremental, $n_2 = 1$. For the iterative formulas when $\mathcal{Q}_2 = \{(t_n, l_n)\}$ Eq. (12) can be simplified to

$$ACS_{d,\mathcal{Q}} = ACS_{d,\mathcal{Q}_1} + CS_{d,\mathcal{Q}_1}\left(-\frac{\Delta_l}{n}\right)$$

$$+ \sum_{p=1}^{d-1}\binom{d}{p}\left(-\frac{\Delta_t}{n}\right)^p\left[ACS_{d-p,\mathcal{Q}_1} + CS_{d-p,\mathcal{Q}_1}\left(-\frac{\Delta_l}{n}\right)\right]$$

$$+ \frac{(-1)^{d+1}(n-1) + (n-1)^{d+1}}{n^{d+1}}(\Delta_t)^d\,\Delta_l, \tag{13}$$

with $\Delta_t = t_n - \mu_{t,\mathcal{Q}_1}$ and $\Delta_l = l_n - \mu_{l,\mathcal{Q}_1}$.

3.3   Denominator

The denominator of Eq. (10) requires the computation of two centralized sums. For the second centralized sum $\sum_{i=1}^{n}(l_i - \mu_l)^2$ we already gave pair-wise iterative as well as incremental formulas for $CS_{2,\mathcal{Q}}$ in Eqs. (6) and (8).

The first centralized sum $\sum_{i=1}^{n}(t'_i - \mu_{t'})^2$ relates to the preprocessed traces. For this, efficient formulas to compute the variance of the preprocessed traces are given in [20]. In order to estimate the variance (second centralized moment $CM_{2,\mathcal{T}'}$) of $\mathcal{T}' = \{t'_{i \in \{1,\dots,n\}}\}$ as the set of preprocessed traces at any arbitrary order $d > 1$ we can write [20]

$$\frac{1}{n}\sum_{i=1}^{n}(t'_i - \mu_{t'})^2 = CM_{2,\mathcal{T}'} = CM_{2d,\mathcal{T}} - (CM_{d,\mathcal{T}})^2 = \frac{CS_{2d,\mathcal{T}}}{n} - \left(\frac{CS_{d,\mathcal{T}}}{n}\right)^2,$$

where $\mathcal{T}$ denotes the traces without preprocessing. Therefore, given the iterative and incremental formulas for $CS_{d,\mathcal{Q}}$ in Eqs. (6) and (8) we can efficiently as well as in parallel estimate both centralized sums of the denominator of Eq. (10). Further, having the formulas given in Sect. 3.2 the correlation of a univariate CPA at any arbitrary order d can be easily derived.

4 Multivariate CPA

In the following we give iterative formula for multivariate higher-order CPA with the optimum combination function, i.e., centered product [16,21]. Given d sample point indices $\mathcal{J} = \{j_1, ..., j_d\}$ as the points to be combined and a set of sample vectors $\mathcal{Q} = \{\, V_{i \in \{1,...,n\}}\,\}$ with $V_i = \left(t_i^{(j)} \mid j \in \mathcal{J}\right)$, the centered product of the ith trace is defined as

$$c_i = \prod_{j \in \mathcal{J}} \left(t_i^{(j)} - \mu_{\mathcal{Q}}^{(j)}\right), \tag{14}$$

where $\mu_{\mathcal{Q}}^{(j)}$ denotes the mean at sample point j over set $\mathcal{Q}$.

The authors of [3] proposed an iterative formula for the Pearson correlation coefficient in the bivariate case, i.e., $d = 2$. However, during the computation they calculate the sum $\sum_{i=1}^{n} \left(t_i^{(j_1)} t_i^{(j_2)}\right)^2$ for the two point indices j_1 and j_2 (cf. s_{11} of Table 5 in [3]). Their method is basically equivalent to using the raw moments to derive higher-order statistical moments. Given a high number of traces this value can grow very large, and can cause numerical instability.

We instead provide iterative formulas based on mean-free values. In our approach, the formula for the multivariate Pearson correlation coefficient is first simplified using Eq. (10) to

$$\rho = \frac{\dfrac{1}{n}\sum_{i=1}^{n}(c_i - \mu_c)(l_i - \mu_l)}{\sqrt{\dfrac{1}{n}\sum_{i=1}^{n}(c_i - \mu_c)^2 \dfrac{1}{n}\sum_{i=1}^{n}(l_i - \mu_l)^2}} = \frac{\dfrac{1}{n}\sum_{i=1}^{n}\left(c_i(l_i - \mu_l)\right)}{\sqrt{\dfrac{1}{n}\sum_{i=1}^{n}(c_i - \mu_c)^2 \dfrac{1}{n}\sum_{i=1}^{n}(l_i - \mu_l)^2}}. \tag{15}$$

4.1 Numerator

The way of computing the numerator of Eq. (15)

$$\frac{1}{n}\sum_{i=1}^{n}\left(c_i(l_i - \mu_l)\right) = \frac{1}{n}\sum_{i=1}^{n}\left(\prod_{j \in \mathcal{J}}\left(t_i^{(j)} - \mu_{\mathcal{Q}}^{(j)}\right)(l_i - \mu_l)\right) \tag{16}$$

is similar to the iterative computation of the first parameter for the multivariate t-test as presented in [20]. We indeed can write Eq. (16) as

$$\frac{1}{n}\sum_{i=1}^{n}\left(c_i(l_i - \mu_l)\right) = \frac{1}{n}\sum_{i=1}^{n}\prod_{j \in \mathcal{J}'}\left(t_i^{(j)} - \mu_{\mathcal{Q}}^{(j)}\right), \tag{17}$$

with $\mathcal{J}' = \mathcal{J} \cup \{j^*\}$, $t_i^{(j*)} = l_i$ and $\mu_{\mathcal{Q}}^{(j*)} = \mu_l$. With this, we define the term *sum of centered products* as

$$SCP_{d+1,\mathcal{Q},\mathcal{J}'} = \sum_{V_i \in \mathcal{Q}} \prod_{j \in \mathcal{J}'} \left(t_i^{(j)} - \mu_{\mathcal{Q}}^{(j)} \right). \tag{18}$$

In addition, we define the b-th order power set of $\mathcal{J}'$ as

$$\mathcal{P}_b = \{\mathcal{S} \mid \mathcal{S} \in \mathbb{P}(\mathcal{J}'), |\mathcal{S}| = b\}, \tag{19}$$

where $\mathbb{P}(\mathcal{J}')$ refers to the power set of the indices of the points of interest $\mathcal{J}'$. The given formulas in [20] are for the incremental case when set $\mathcal{Q}_2$ has a cardinality of 1. Hence, the sum of the centered products $SCP_{d+1,\mathcal{Q},\mathcal{J}'}$ of the extended set $\mathcal{Q} = \mathcal{Q}_1 \cup \left\{ (t_n^{(j_1)}, ..., t_n^{(j_d)}, t_n^{(j^*)}) \right\}$ with $t_n^{(j^*)} = l_n$ and $|\mathcal{Q}| = n$ can be computed as [20]

$$SCP_{d+1,\mathcal{Q},\mathcal{J}'} = SCP_{d+1,\mathcal{Q}_1,\mathcal{J}'} + \left(\sum_{b=2}^{d} \sum_{\mathcal{S} \in \mathcal{P}_b} SCP_{b,\mathcal{Q}_1,\mathcal{S}} \prod_{j \in \mathcal{J}'\backslash \mathcal{S}} \left(\frac{\Delta^{(j)}}{-n} \right) \right)$$
$$+ \left(\frac{(-1)^{d+1}(n-1) + (n-1)^{d+1}}{n^{d+1}} \prod_{j \in \mathcal{J}'} \Delta^{(j)} \right), \tag{20}$$

where $\Delta^{(j \in \mathcal{J}')} = t_n^{(j)} - \mu_{\mathcal{Q}_1}^{(j)}$. Below we present a generalization of this method to arbitrary sized $\mathcal{Q}_2$.

Generalization of [20]

Theorem 2. *Let $\mathcal{J}'$ be a given set of indices (of $d + 1$ points of interest) and two sets of sample vectors $\mathcal{Q}_1 = \{V_{i \in \{1,...,n_1\}}\}$, $\mathcal{Q}_2 = \{V_{i \in \{1,...,n_2\}}\}$ with $V_i = \left(t_i^{(j)} \mid j \in \mathcal{J}' \right)$. The sum of the centered products $SCP_{d+1,\mathcal{Q},\mathcal{J}'}$ of the extended set $\mathcal{Q} = \mathcal{Q}_1 \cup \mathcal{Q}_2$ with $\Delta^{(j \in \mathcal{J}')} = \mu_{\mathcal{Q}_2}^{(j)} - \mu_{\mathcal{Q}_1}^{(j)}$ and $|\mathcal{Q}| = n$ can be computed as:*

$$SCP_{d+1,\mathcal{Q},\mathcal{J}'} = SCP_{d+1,\mathcal{Q}_1,\mathcal{J}'} + SCP_{d+1,\mathcal{Q}_2,\mathcal{J}'}$$
$$+ \sum_{b=2}^{d} \sum_{\mathcal{S} \in \mathcal{P}_b} \left((-n_2)^{d+1-b} SCP_{b,\mathcal{Q}_1,\mathcal{S}} + n_1^{d+1-b} SCP_{b,\mathcal{Q}_2,\mathcal{S}} \right) \prod_{j \in \mathcal{J}'\backslash \mathcal{S}} \frac{\Delta^{(j)}}{n}$$
$$+ \frac{(-n_2)^{d+1} n_1 + n_1^{d+1} n_2}{n^{d+1}} \prod_{j \in \mathcal{J}'} \Delta^{(j)}. \tag{21}$$

The proof of Theorem 2 is omitted due to length restrictions.

4.2 Denominator

Similar to the expressions given in Sect. 3.3 the denominator of Eq. (15) consists of two centralized sums. The second one $\sum_{i=1}^{n} (l_i - \mu_l)^2$ is the same as that of the univariate CPA and Eqs. (6) and (8) are still valid.

For the first centralized sum $\sum_{i=1}^{n} (c_i - \mu_c)^2$ we recall the formulas given in [20] which deal with the estimation of the variance of the preprocessed traces in a multivariate setting. It means that we can write

$$\sum_{i=1}^{n} \left(c_i - \mu_c\right)^2 = \sum_{V \in \mathcal{Q}} \left(\prod_{j \in \mathcal{J}} \left(t^{(j)} - \mu_{\mathcal{Q}}^{(j)}\right) - \frac{SCP_{d,\mathcal{Q},\mathcal{J}}}{n} \right)^2$$

$$= SCP_{2d,\mathcal{Q},\mathcal{J}''} - \frac{(SCP_{d,\mathcal{Q},\mathcal{J}})^2}{n}, \tag{22}$$

with multiset $\mathcal{J}'' = \{j_1, ..., j_d, j_1, ..., j_d\}$. It is noteworthy that in contrast to the computation of the numerator, where the set $\mathcal{J}'$ with $d+1$ indices is used, here for the denominator the set $\mathcal{J}$ and its extension $\mathcal{J}''$ with respectively d and $2d$ indices are applied.

5 Moments-Correlating DPA

Moments-Correlating DPA (MC-DPA) [14] as a successor of Correlation-Enhanced Power Analysis Collision Attack [12] solves its shortcomings and is based on correlating the moments to the traces [7,8,11]. It relaxes the necessity of a hypothetical leakage model which is essential in the case of a CPA.

The most general form of MC-DPA is Moments-Correlating Profiling DPA (MCP-DPA). In such a scenario, the traces used to build the model $t_{i\in\{1,...,n^{(M)}\}}^{(M)}$ (and trivially their number $n^{(M)}$) are not necessarily the same as the traces used in the attack $t_{i\in\{1,...,n\}}$. An MC-DPA in a multivariate settings uses two sets of sample point indices $\mathcal{J}_M$ and $\mathcal{J}_t$ related to the sample points of the model and the attack respectively. Such sample points are taken based on the time instances when a certain function (e.g., an Sbox) operates on an intermediate value $v_{i\in\{1,...,n^{(M)}\}}^{(M)}$ to form the model and on another intermediate value $v_{i\in\{1,...,n\}}^{(t)}$ to perform the attack. In a simple scenario, such intermediate values can be different Sbox inputs. Optionally a leakage function can be considered as $\widetilde{L}(.)$ over the targeted intermediate values. Note that in the most general form such a leakage function can be the identity mapping, i.e., $\widetilde{L}(v) = v$. Following the original MC-DPA scheme [14], $v_i^{(M)} = d_i^{(M)} \oplus k^{(M)}$ and $v_i^{(t)} = d_i^{(t)} \oplus k^{(t)}$ with $d^{(M)}$ and $d^{(t)}$ e.g., plaintext portions (bytes) respectively of the model and the attack. Hence, due to the linear relations such a setting turns into a linear collision attack [2] with $\widetilde{L}(v_i^{(M)}) = d_i^{(M)}$ and $\widetilde{L}(v_i^{(t)}) = d_i^{(t)} \oplus \Delta k$, which is referred to as Moments-Correlating Collision DPA (MCC-DPA), where the traces for the model and the attack are the same and $n^{(M)} = n$. However, in the following expressions we consider the profiling one which can be easily simplified to the collision one.

Let us denote $\mathcal{L}$ as a set of all possible outputs of the leakage function with cardinality of $n_{\mathcal{L}}$ is defined as

$$\mathcal{L} = \{l^{(1)}, \dots, l^{(n_{\mathcal{L}})}\} = \{l \mid \exists v, \widetilde{\mathsf{L}}(v) = l\}. \tag{23}$$

Correspondingly we define $n_{\mathcal{L}}$ subsets $\mathcal{I}_{l^{(a \in \{1, \dots, n_{\mathcal{L}}\})}}^{(M)}$

$$\mathcal{I}_{l^{(a)}}^{(M)} = \{i \in \{1, \dots, n^{(M)}\} \mid \widetilde{\mathsf{L}}(v_i^{(M)}) = l^{(a)}\} \tag{24}$$

as the trace indices with particular leakage value $l^{(a)}$ on the model's intermediate values $v_i^{(M)}$ with cardinality of $n_{l^{(a)}}^{(M)}$. The same subsets are also defined with respect to the attack's intermediate values $v_i^{(t)}$ as

$$\mathcal{I}_{l^{(a)}}^{(t)} = \{i \in \{1, \dots, n\} \mid \widetilde{\mathsf{L}}(v_i^{(t)}) = l^{(a)}\}, \tag{25}$$

with $|\mathcal{I}_{l^{(a)}}^{(t)}| = n_{l^{(a)}}^{(t)}$.

Depending on the type of the attack (univariate vs. multivariate) the sample points at $\mathcal{J}_M$ are first combined using a combining function, e.g., centered product, split into the subsets depending the leakage model $\widetilde{\mathsf{L}}(.)$ and then used to estimate the statistical moments of a given order d. Depending on the order of the attack, prior preprocessing is also necessary. We denote these moments as the model by

$$\forall l^{(a)} \in \mathcal{L}, \; M_{l^{(a)}} \xleftarrow[\text{dth-order moment}]{\substack{\text{preprocessing,} \\ \text{(centralized/standardized)}}} \{t_i^{(M)}, i \in \mathcal{I}_{l^{(a)}}^{(M)}, \mathcal{J}_M\}. \tag{26}$$

On the other hand, the traces at the sample points $\mathcal{J}_t$ need also to be preprocessed according to the variate of the attack (univariate vs. multivariate) as well as the given order d.

The correlation between the moments $M_{l^{(a \in \{1, \dots, n_{\mathcal{L}}\})}}$ and the preprocessed traces $t'_{i \in \{1, \dots, n\}}$ is defined as

$$\rho = \frac{\dfrac{1}{n} \sum_{i=1}^{n} (t'_i - \mu_{t'})(M_{l_i} - \mu_M)}{\sqrt{\dfrac{1}{n} \sum_{i=1}^{n} (t'_i - \mu_{t'})^2 \dfrac{1}{n} \sum_{i=1}^{n} (M_{l_i} - \mu_M)^2}}, \tag{27}$$

where $M_{l_{i \in \{1, \dots, n\}}} = M_{l^{(a)}}, \; l^{(a)} = \widetilde{\mathsf{L}}(v_i^{(t)}) \in \mathcal{L}$.

5.1 Numerator

To compute the numerator of Eq. (27) it is first simplified to

$$\frac{1}{n}\sum_{i=1}^{n}(t'_i - \mu_{t'})(M_{l_i} - \mu_M) = \sum_{a=1}^{n_{\mathcal{L}}}(M_{l^{(a)}} - \mu_M)\frac{1}{n}\sum_{i\in\mathcal{I}_{l^{(a)}}^{(t)}} t'_i. \tag{28}$$

The preprocessing of the MC-DPA requires the sum of Eq. (28) $SUM_{\mathcal{I}_{l^{(a)}}^{(t)}} = \sum_{i\in\mathcal{I}_{l^{(a)}}^{(t)}} t'_i$ to be processed independently. Otherwise, it is not trivially possible to provide iterative formulas as the mean and variance of subgroup of the traces $\in \mathcal{I}_{l^{(a)}}^{(t)}$ change. Since $n_{\mathcal{L}}$ is limited, we store a sum for each value of set $\mathcal{L}$ and merge them only at the end when the value of the estimated correlation is desired. In the multivariate higher-order $d > 1$ scenario, we store $n_{\mathcal{L}}$ sums of the traces as

$$SUM_{\mathcal{I}_{l^{(a)}}^{(t)}} = \sum_{i\in\mathcal{I}_{l^{(a)}}^{(t)}} t'_i = \sum_{i\in\mathcal{I}_{l^{(a)}}^{(t)}} \prod_{j\in\mathcal{J}_t}\left(t_i^{(j)} - \mu_{\mathcal{I}_{l^{(a)}}^{(t)}}^{(j)}\right) = SCP_{d,\mathcal{I}_{l^{(a)}}^{(t)},\mathcal{J}_t}, \tag{29}$$

and in case of the univariate higher-order $d > 2$ as

$$SUM_{\mathcal{I}_{l^{(a)}}^{(t)}} = \sum_{i\in\mathcal{I}_{l^{(a)}}^{(t)}} t'_i = \frac{1}{\left(s_{\mathcal{I}_{l^{(a)}}^{(t)}}\right)^d}\sum_{i\in\mathcal{I}_{l^{(a)}}^{(t)}}\left(t_i - \mu_{\mathcal{I}_{l^{(i)}}^{(t)}}\right)^d = \frac{1}{\left(s_{\mathcal{I}_{l^{(i)}}^{(t)}}\right)^d}CS_{d,\mathcal{I}_{l^{(i)}}^{(t)}}. \tag{30}$$

Note that for $d = 2$ the denominator of Eq. (30) is omitted. For a univariate first-order attack the means are used to derive the latter term of Eq. (28) as

$$\frac{1}{n}SUM_{\mathcal{I}_{l^{(a)}}^{(t)}} = \frac{1}{n}\sum_{i\in\mathcal{I}_{l^{(a)}}^{(t)}} t_i = \frac{n_{l^{(a)}}^{(t)}}{n}\mu_{\mathcal{I}_{l^{(a)}}^{(t)}}. \tag{31}$$

We should here emphasize that – in contrast to the methods of the prior sections – in case of MC-DPA when a new trace is added to the set of traces following the incremental formulas only the sum and the moments which correspond to the leakage value $l^{(a)}$ related to the new trace are updated.

In order to calculate the whole numerator it is necessary to store the moments $M_{l^{(a)}}, \forall l^{(a)} \in \mathcal{L}$. This procedure is similar to before, and for the multivariate higher-order case it can be done by computing

$$M_{l^{(a)}} = \frac{1}{n_{l^{(a)}}^{(M)}}\sum_{i\in\mathcal{I}_{l^{(a)}}^{(M)}} \prod_{j\in\mathcal{J}_M}\left(t_i^{(j)} - \mu_{\mathcal{I}_{l^{(a)}}^{(M)}}^{(j)}\right) = \frac{SCP_{d,\mathcal{I}_{l^{(a)}}^{(M)},\mathcal{J}_M}}{n_{l^{(a)}}^{(M)}}. \tag{32}$$

For the univariate case Eq. (32) changes analog Eq. (30). In a univariate first-order attack there is no preprocessing, and $M_{l^{(a)}}$ simply represents the mean $\mu_{\mathcal{I}_{l^{(a)}}^{(M)}}$.

The mean μ_M in Eq. (27) is

$$\mu_M = \frac{1}{n} \sum_{a=1}^{n_{\mathcal{L}}} n_{l(a)}^{(t)} M_{l(a)}, \tag{33}$$

and as an example in case of a multivariate higher-order attack can be written as

$$\mu_M = \frac{1}{n} \sum_{a=1}^{n_{\mathcal{L}}} SCP_{d, \mathcal{I}_{l(i)}^{(t)}, \mathcal{J}_M}. \tag{34}$$

Since the iterative formulas (for both pair-wise and incremental cases) to compute $SCP_{d,\dots}$ and $CS_{d,\dots}$ as well as other necessary moments are given in previous sections, the numerator of Eq. (27) can be easily derived.

5.2   Denominator

The first part of the denominator can be written as

$$\frac{1}{n} \sum_{i=1}^{n} (t_i' - \mu_{t'})^2 = \frac{1}{n} \sum_{i=1}^{n} t_i'^2 - (\mu_{t'})^2 = \frac{1}{n} \sum_{a=1}^{n_{\mathcal{L}}} \left(\sum_{i \in \mathcal{I}_{l(a)}^{(t)}} t_i'^2 \right) - (\mu_{t'})^2. \tag{35}$$

Therefore, we additionally need to compute the sums of the squared preprocessed traces $SUM_{\mathcal{I}_{l(a)}^{(t)}}^2 = \sum_{i \in \mathcal{I}_{l(a)}^{(t)}} t_i'^2$. For a multivariate higher-order case, this can be written as $SCP_{2d, \mathcal{I}_{l(a)}^{(t)}, \{\mathcal{J}_t, \mathcal{J}_t\}}$ similar to Eq. (29) or similar to Eqs. (30) and (31) for the univariate cases. Further, the sums $SUM_{\mathcal{I}_{l(a)}^{(t)}}$ computed by Eqs. (29) and (30), or Eq. (31) can be used to derive $\mu_{t'}$ following the same principle of Eq. (33).

The second part of the denominator of Eq. (27) can be obtained from the values that are already used to compute the numerator:

$$\frac{1}{n} \sum_{i=1}^{n} (M_{l_i} - \mu_M)^2 = \frac{1}{n} \sum_{a=1}^{n_{\mathcal{L}}} n_{l(a)}^{(t)} (M_{l(a)} - \mu_M)^2. \tag{36}$$

Since $n_{\mathcal{L}}$ is limited, the above expression can be computed at the end when all traces are processed to estimate the correlation.

In the aforementioned approach the sums $SUM_{\mathcal{I}_{l(a)}^{(t)}}$ are grouped based on the output of the leakage function, i.e., $l^{(a)}$, which is also key dependent. Hence, the traces have to be regrouped for each key candidate as well as for each selected leakage function $\widetilde{L}(.)$.

6   Evaluation

We evaluate the accuracy (convergence) of our presented approaches, and compare it to the corresponding results of the raw-moment and three-pass

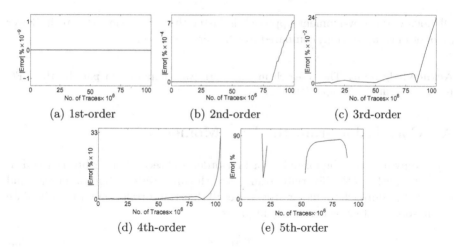

Fig. 1. Difference between the result of correlation estimations (raw-moment versus three-pass)

approaches. To this end, we generate 100 million simulated leakages by $\sim\mathcal{N}(100+\mathrm{HW}(x),3)$, where x is drawn uniformly from $\{0,1\}^4$. Hence, the correlation between the leakages and $\mathrm{HW}(x)$ is estimated. Following the concept of higher-order attacks, the leakages are also preprocessed (up to fifth order) to allow an emulation of a higher-order univariate CPA. Note that the performance results are still valid in the multivariate case given additional leakage points with a similar leakage structure and the normalized product as combination function. This can be easily seen as both type of attacks require the estimation of centralized values up to a power of $2d$ (with an additional standardization for univariate higher-order attacks). The results based on our incremental approaches are exactly the same to the three-pass ones, i.e., with absolute 0 difference. As [3] only includes the formulas for first-order and second-order bivariate CPA, we further had to derive the necessary formulas for the univariate correlation up to the fifth order. The formulas can be found in Appendix A.

With these formulas we computed the correlation up to the fifth order on an Intel Xeon X5670 using a single thread, and examined the differences with respect to the results of the three-pass approach. Figure 1 presents the corresponding results. As expected, in the first-order setting the results are exactly the same, but the differences start to be obvious at higher orders particularly for higher number of traces. It is noteworthy that in the cases where no difference is shown for the fifth-order correlation, one of the variances of the denominator in the raw-moment approach turned to a negative value which indicates the instability of such formulas. With respect to the execution time of each approach, although it depends on the optimization level of the underlying computer code, we report 43 s, 17.8 s, and 11.6 s for three-pass, our incremental, and raw-moment approach respectively to estimate all five correlations at the same time on 100 million leakage points.

Obviously, the raw-moment approach is faster than the others due to its lower amount of computations compared to our incremental one.

Acknowledgment. The research in this work was supported in part by the DFG Research Training Group GRK 1817/1.

A   Correlation from the Raw Moments

As [3] only includes the formulas for first-order and second-order bivariate CPA, we first transform the bivariate formulas to the univariate second-order case and extend the approach to higher orders. Recall that the correlation for the bivariate second-order attack is computed in [3] as

$$\rho = \frac{n\lambda_1 - \lambda_2 s_3}{\sqrt{n\lambda_3 - \lambda_2^2}\sqrt{ns_9 - s_3^2}}, \tag{37}$$

where n denotes the number of traces and $\lambda_{\{1,2,3\}}$ are derived from the sums $s_{\{1,\dots,13\}}$.

For the univariate second-order correlation, some of these sums are equivalent. Therefore, in this special case it is possible to reduce the number of sums required to be computed. For that, we first denote the d-th order sums as

$$S_d^{(t)} = \sum_{i=1}^n t_i^d, \qquad S_d^{(l)} = \sum_{i=1}^n l_i^d, \qquad S_d^{(t,l)} = \sum_{i=1}^n t_i^d l \tag{38}$$

with $s_3 = S_1^{(l)}$ and $s_9 = S_2^{(l)}$. The remaining parameters are then derived as

$$\lambda_1 = S_2^{(t,l)} - 2\frac{S_1^{(t)} S_1^{(t,l)}}{n} + \frac{S_1^{(t)} S_1^{(t)} S_1^{(l)}}{n^2}, \quad \lambda_2 = S_2^{(t)} - \frac{S_1^{(t)} S_1^{(t)}}{n}, \tag{39}$$

$$\lambda_3 = S_4^{(t)} - 4\frac{S_1^{(t)} S_3^{(t)}}{n} + 6\frac{S_1^{(t)} S_1^{(t)} S_2^{(t)}}{n^2} - 3\frac{S_1^{(t)} S_1^{(t)} S_1^{(t)} S_1^{(t)}}{n^3}. \tag{40}$$

For the higher-order correlation the basic structure of Eq. (37) stays the same, and only the formulas for $\lambda_{\{1,2,3\}}$ change. We provided all necessary formulas in the following subsections.

A.1   Third Order

$$\lambda_1 = S_3^{(t,l)} - 3\frac{S_1^{(t)} S_2^{(t,l)}}{n} + 3\frac{\left(S_1^{(t)}\right)^2 S_1^{(t,l)}}{n^2} - \frac{\left(S_1^{(t)}\right)^3 S_1^{(l)}}{n^3}, \tag{41}$$

$$\lambda_2 = S_3^{(t)} - 3\frac{S_1^{(t)} S_2^{(t)}}{n} + 2\frac{\left(S_1^{(t)}\right)^3}{n^2}, \tag{42}$$

$$\lambda_3 = S_6^{(t)} - 6\frac{S_1^{(t)} S_5^{(t)}}{n} + 15\frac{\left(S_1^{(t)}\right)^2 S_4^{(t)}}{n^2} - 20\frac{\left(S_1^{(t)}\right)^3 S_3^{(t)}}{n^3}$$

$$+ 15\frac{\left(S_1^{(t)}\right)^4 S_2^{(t)}}{n^4} - 5\frac{\left(S_1^{(t)}\right)^6}{n^5} \tag{43}$$

A.2   Fourth Order

$$\lambda_1 = S_4^{(t,l)} - 4\frac{S_1^{(t)} S_3^{(t,l)}}{n} + 6\frac{\left(S_1^{(t)}\right)^2 S_2^{(t,l)}}{n^2} - 4\frac{\left(S_1^{(t)}\right)^3 S_1^{(t,l)}}{n^3} + \frac{\left(S_1^{(t)}\right)^4 S_1^{(l)}}{n^4}, \tag{44}$$

$$\lambda_2 = S_4^{(t)} - 4\frac{S_1^{(t)} S_3^{(t)}}{n} + 6\frac{\left(S_1^{(t)}\right)^2 S_2^{(t)}}{n^2} - 3\frac{\left(S_1^{(t)}\right)^4}{n^3}, \tag{45}$$

$$\lambda_3 = S_8^{(t)} - 8\frac{S_1^{(t)} S_7^{(t)}}{n} + 28\frac{\left(S_1^{(t)}\right)^2 S_6^{(t)}}{n^2} - 56\frac{\left(S_1^{(t)}\right)^3 S_5^{(t)}}{n^3}$$

$$+ 70\frac{\left(S_1^{(t)}\right)^4 S_4^{(t)}}{n^4} - 56\frac{\left(S_1^{(t)}\right)^5 S_3^{(t)}}{n^5} + 28\frac{\left(S_1^{(t)}\right)^6 S_2^{(t)}}{n^6} - 7\frac{\left(S_1^{(t)}\right)^8}{n^7} \tag{46}$$

A.3 Fifth Order

$$\lambda_1 = S_5^{(t,l)} - 5\frac{S_1^{(t)}S_4^{(t,l)}}{n} + 10\frac{\left(S_1^{(t)}\right)^2 S_3^{(t,l)}}{n^2} - 10\frac{\left(S_1^{(t)}\right)^3 S_2^{(t,l)}}{n^3}$$

$$+ 5\frac{\left(S_1^{(t)}\right)^4 S_1^{(t,l)}}{n^4} - \frac{\left(S_1^{(t)}\right)^5 S_1^{(l)}}{n^5}, \tag{47}$$

$$\lambda_2 = S_5^{(t)} - 5\frac{S_1^{(t)}S_4^{(t)}}{n} + 10\frac{\left(S_1^{(t)}\right)^2 S_3^{(t)}}{n^2} - 10\frac{\left(S_1^{(t)}\right)^3 S_2^{(t)}}{n^3} + 4\frac{\left(S_1^{(t)}\right)^5}{n^4}, \tag{48}$$

$$\lambda_3 = S_{10}^{(t)} - 10\frac{S_1^{(t)}S_9^{(t)}}{n} + 45\frac{\left(S_1^{(t)}\right)^2 S_8^{(t)}}{n^2} - 120\frac{\left(S_1^{(t)}\right)^3 S_7^{(t)}}{n^3} + 210\frac{\left(S_1^{(t)}\right)^4 S_6^{(t)}}{n^4}$$

$$- 252\frac{\left(S_1^{(t)}\right)^5 S_5^{(t)}}{n^5} + 210\frac{\left(S_1^{(t)}\right)^6 S_4^{(t)}}{n^6} - 120\frac{\left(S_1^{(t)}\right)^7 S_3^{(t)}}{n^7} + 45\frac{\left(S_1^{(t)}\right)^8 S_2^{(t)}}{n^8}$$

$$- 9\frac{\left(S_1^{(t)}\right)^{10}}{n^9} \tag{49}$$

References

1. Bilgin, B., Gierlichs, B., Nikova, S., Nikov, V., Rijmen, V.: Higher-order threshold implementations. In: Sarkar, P., Iwata, T. (eds.) ASIACRYPT 2014, Part II. LNCS, vol. 8874, pp. 326–343. Springer, Heidelberg (2014)
2. Bogdanov, A.: Multiple-differential side-channel collision attacks on AES. In: Oswald, E., Rohatgi, P. (eds.) CHES 2008. LNCS, vol. 5154, pp. 30–44. Springer, Heidelberg (2008)
3. Bottinelli, P., Bos, J.W.: Computational Aspects of Correlation Power Analysis. Cryptology ePrint Archive, Report 2015/260 (2015). http://eprint.iacr.org/
4. Brier, E., Clavier, C., Olivier, F.: Correlation power analysis with a leakage model. In: Joye, M., Quisquater, J.-J. (eds.) CHES 2004. LNCS, vol. 3156, pp. 16–29. Springer, Heidelberg (2004)
5. Chari, S., Jutla, C.S., Rao, J.R., Rohatgi, P.: Towards sound approaches to counteract power-analysis attacks. In: Wiener, M. (ed.) CRYPTO 1999. LNCS, vol. 1666, p. 398. Springer, Heidelberg (1999)
6. Duc, A., Dziembowski, S., Faust, S.: Unifying leakage models: from probing attacks to noisy leakage. In: Nguyen, P.Q., Oswald, E. (eds.) EUROCRYPT 2014. LNCS, vol. 8441, pp. 423–440. Springer, Heidelberg (2014)
7. Duc, A., Faust, S., Standaert, F.-X.: Making masking security proofs concrete. In: Oswald, E., Fischlin, M. (eds.) EUROCRYPT 2015. LNCS, vol. 9056, pp. 401–429. Springer, Heidelberg (2015)
8. Durvaux, F., Standaert, F.-X., Veyrat-Charvillon, N., Mairy, J.-B., Deville, Y.: Efficient selection of time samples for higher-order DPA with projection pursuits. In: Mangard, S., Poschmann, A.Y. (eds.) COSADE 2015. LNCS, vol. 9064, pp. 34–50. Springer, Heidelberg (2015)

9. Goodwill, G., Jun, B., Jaffe, J., Rohatgi, P.: A testing methodology for side channel resistance validation. In: NIST Non-invasive Attack Testing Workshop (2011). http://csrc.nist.gov/news_events/non-invasive-attack-testing-workshop/papers/08_Goodwill.pdf

10. Higham, N.J.: Accuracy and Stability of Numerical Algorithms, 2nd edn. SIAM, Philadelphia (2002)

11. Moradi, A., Immler, V.: Early propagation and imbalanced routing, how to diminish in FPGAs. In: Batina, L., Robshaw, M. (eds.) CHES 2014. LNCS, vol. 8731, pp. 598–615. Springer, Heidelberg (2014)

12. Moradi, A., Mischke, O., Eisenbarth, T.: Correlation-enhanced power analysis collision attack. In: Mangard, S., Standaert, F.-X. (eds.) CHES 2010. LNCS, vol. 6225, pp. 125–139. Springer, Heidelberg (2010)

13. Moradi, A., Poschmann, A., Ling, S., Paar, C., Wang, H.: Pushing the limits: a very compact and a threshold implementation of AES. In: Paterson, K.G. (ed.) EUROCRYPT 2011. LNCS, vol. 6632, pp. 69–88. Springer, Heidelberg (2011)

14. Moradi, A., Standaert, F.: Moments-Correlating DPA. Cryptology ePrint Archive, Report 2014/409 (2014). http://eprint.iacr.org/

15. Pébay, P.: Formulas for Robust, One-Pass Parallel Computation of Covariances and Arbitrary-Order Statistical Moments. Sandia Report SAND-6212, Sandia National Laboratories (2008)

16. Prouff, E., Rivain, M., Bevan, R.: Statistical analysis of second order differential power analysis. IEEE Trans. Comput. 58(6), 799–811 (2009)

17. Rao, J.R., Rohatgi, P., Scherzer, H., Tinguely, S., Attacks, P.: Or How to rapidly clone some GSM cards. In: IEEE Symposium on Security and Privacy, pp. 31–41. IEEE Computer Society (2002)

18. Reparaz, O., Gierlichs, B., Verbauwhede, I.: Selecting time samples for multivariate DPA attacks. In: Prouff, E., Schaumont, P. (eds.) CHES 2012. LNCS, vol. 7428, pp. 155–174. Springer, Heidelberg (2012)

19. Rivain, M., Prouff, E.: Provably secure higher-order masking of AES. In: Mangard, S., Standaert, F.-X. (eds.) CHES 2010. LNCS, vol. 6225, pp. 413–427. Springer, Heidelberg (2010)

20. Schneider, T., Moradi, A.: Leakage assessment methodology. In: Güneysu, T., Handschuh, H. (eds.) CHES 2015. LNCS, vol. 9293, pp. 495–513. Springer, Heidelberg (2015)

21. Standaert, F.-X., Veyrat-Charvillon, N., Oswald, E., Gierlichs, B., Medwed, M., Kasper, M., Mangard, S.: The world is not enough: another look on second-order DPA. In: Abe, M. (ed.) ASIACRYPT 2010. LNCS, vol. 6477, pp. 112–129. Springer, Heidelberg (2010)

22. Zhou, Y., Yu, Y., Standaert, F.-X., Quisquater, J.-J.: On the need of physical security for small embedded devices: a case study with COMP128-1 implementations in SIM cards. In: Sadeghi, A.-R. (ed.) FC 2013. LNCS, vol. 7859, pp. 230–238. Springer, Heidelberg (2013)

Author Index

Printed in the United States
By Bookmasters